Modern Security Methods

Modern Security Methods

Charles F. Hemphill, Jr.
M. S., Doctor of Jurisprudence
*Member American Society
for Industrial Security*
*Special Consultant,
Wells Fargo Guard Services*

PRENTICE-HALL, INC., Englewood Cliffs, New Jersey 07632

Library of Congress Cataloging in Publication Data

HEMPHILL, CHARLES F
   Modern security methods.

   Includes bibliographical references and index.
   1. Industry—Security measures. 2. Retail trade—
Security measures. I. Title.
HV8290.H44   658.4'7   78-23459
ISBN 0-13-069203-4

Editorial/production supervision and interior
   design by Dee Amir Josephson
Cover design by Saiki/Sprung Design
Manufacturing buyer: John Hall

© 1979 by Prentice-Hall, Inc., Englewood Cliffs, N.J. 07632
All rights reserved. No part of this book
may be reproduced in any form or
by any means without permission in writing
from the publisher.

Printed in the United States of America

10 9 8 7 6 5 4 3 2 1

PRENTICE-HALL INTERNATIONAL, INC., *London*
PRENTICE-HALL of Australia Pty. Limited, *Sydney*
PRENTICE-HALL of Canada, Ltd., *Toronto*
PRENTICE-HALL of India Private Limited, *New Delhi*
PRENTICE-HALL of Japan, Inc., *Tokyo*
PRENTICE-HALL of Southeast Asia Pte. Ltd., *Singapore*
WHITEHALL BOOKS LIMITED, *Wellington, New Zealand*

Contents

Preface xvii

1 INTRODUCTION 1

Background and General Problems of Business and Industrial Security 1

> *A Definition of Security and What it Implies;*
> *Humanity's Long Struggle to Maintain Security;*
> *Two Basic Kinds of Security;*
> *The Increase in Crime—Its Impact on Business*

High Points in the History and Development of Private Security 4

The Services Provided by Private Security 8

> *Guard Services; Patrol Services; Investigation Services;*
> *Armored Car Service; The Central Alarm Station;*
> *Undercover Operators; Loss Prevention Surveys;*
> *Predeparture Screening; Guard Dogs;*
> *Polygraph Examinations (Lie Detector Tests);*
> *Executive Protection; Shopping Services;*
> *"Sweeps" for Bugging Devices; Safety; Fire Protection;*
> *Psychological Stress Evaluator (PSE)*

Summary 14

Review and Discussion 15

Good General Sources on Security 15

2 PHYSICAL FACILITIES 16

Building Location and Controls 16

*Physical Controls—Deterrents to Entry;
Three Lines of Defense for Physical Security;
Choosing the Neighborhood
and Planning Building Placement; Landscaping;
Planning the Building Proper for Good Security*

Perimeter Controls 18

The Fence as Perimeter Protection

Building Openings 20

*Building Exit and Entry Controls;
Padlocking all Doors Inside Except the Lone Control Door;
Warehouse Doors; Door Hinge Pins; Windows;
Steel Mesh or Iron Bars on Windows;
Using Doors and Windows for Ventilation;
Painting the Inside of Warehouse Windows;
Window or Door Locks that Can be Reached
by Breaking Glass;
The Roof Hatch, Skylight, Air Conditioning Vent;
Basement Access; Ladders; Public Utility Openings*

Summary 32

Review and Discussion 33

3 LOCKS AND KEYS: SAFES AND VAULTS 35

Locks and Keys 35

*Cylinder Pulling; Springing the Door; Jamb Peeling;
Bar and Hammer Attacks, and Sawing the Bolt;
Padlock Substitution; Protecting Hasp Installation;
Removable Core Locks; Specialized Locks; Key Control;
Key Audits as a Control;
Precaution Against Key Duplication;
Security for Duplicate Keys; Maintaining Key Records*

Security Features of Business Safes and Vaults 50

*Selecting a Location for the Money Safe;
Making Certain the Money Safe Is Not Easy to Carry Away;
Controlling Tools and Equipment That May Be Used
Against the Safe; Special Safes for Delivery Vehicles;*

Walk-In Vaults; Using the Protective Features of the Safe; Insulated and Lockable File Cabinets for Security; Lockable Cash Drawers

Summary 57

Review and Discussion 59

4 PROBLEMS OF MAINTENANCE AND JANITORIAL ACCESS 60

Typical Problems 60

Access of Maintenance Employees to Buildings and Valuables; Tools and Equipment; Tool Thefts; Marking the Firm's Tools; Replacement Tools; Searching Employees for Tools; Automobile Supplies and Gasoline; Welding Equipment; Pallets and Hand Trucks; The Forklift as a Security Threat; Company Uniforms; Trash Disposal Problems; Security Personnel Doubling in Safety

Providing Security Through Fire Protection 70

Water Sprinkler Systems; Classes of Fires; Use of Extinguishers or Portable Equipment; Classes of Extinguishers; Placement and Use of Hand Extinguishers; Fire Hoses; Fire Doors; Fire Escapes; Fire Detectors; Additional Guard Activity in Case of Fire

Summary 77

Review and Discussion 79

5 ALARMS AND OTHER PROTECTIVE SYSTEMS 80

Alarms 80

Basic Parts of an Alarm System; Benefits and Uses of an Intrusion Alarm; Assessing the Need for an Alarm System; Different Kinds of Systems That Get the Word to Police or Management

Alarm Sensoring or Triggering Systems 84

Electromechanical Devices—Tape or Foil and Contacts; Photoelectric Alarm Installation; Motion Detection Alarms;

*Sonic Alarm Systems; Vibration Detectors; Pressure Alarms;
Fence Alarms; Capacitance Alarms or Proximity Alarms*

Closed Circuit Television (CCTV) 87

Protection Services 88

*The Security Guard;
Responsibilities and Limitations of Guards; Guard Arrests;
Providing Weapons to Guards;
Contract Guards vs. Company Guards;
Watchclock Systems; Patrol Services; Guard Dogs*

Summary 95

Review and Discussion 96

6 ACCESS CONTROLS 98

The Importance and Use of Access Controls 98

*The Parking Lot; The Patrol Vehicle on the Parking Lot;
Night Lighting; Replacing Fixtures and Lights;
Some Principles of Protective Lighting;
Additional Lighting for Some Areas;
Night Lighting Standards; The Guard Shack;
Closed Circuit TV*

Badge Systems as a Means of Control 104

*The Temporary Badge; Visitor Badges;
Suppliers and Vendors; Package Passes;
Card-Key Access Systems; Company Tours*

Other Weaknesses in Access 112

*The Catering Truck; Employee Lockers;
Restricting Access Inside a Building or Facility*

Summary 113

Review and Discussion 114

7 THEFT 116

Some Background to the Problem 116

*What Is the Extent of Theft in the United States?;
What Are the Real Consequences
of Theft in the Business World?;
Do Thefts Involve Taking of Money or Merchandise?;
Are Most Thefts from Inside or Outside the Business?;*

Contents ix

> *Why Employees Steal?; Some Security
> and Management Controls to Minimize Theft;
> An Example of an Audit*

Kinds of Thefts by Employees 121

> *Employee Appropriation as a Form of Theft;
> Taking Merchandise Samples;
> Theft of Side Products, Scrap, and Salvage;
> Employee Purchases; Theft at the Will Call Counter;
> The "Early Employee"; Merchandise Refunds;
> Driver Collection Losses;
> Ticket Switching and Merchandise Substitution;
> Till Tapping*

Summary 128

Review and Discussion 129

8 SHOPLIFTING 131

Introduction 131

> *Who Shoplifts, for What, and When;
> Designing the Building to Reduce Shoplifting;
> How Shoplifting Is Done; Conditions and Factors Involved;
> Good Customer Service as Major Factor
> in Reducing Shoplifting;
> Techniques for Displaying and Protecting Merchandise;
> When the Sales Clerk Is Not Certain;
> Apprehension and Prosecution;
> Some Problems in Clothing Stores;
> Some Other Protective Techniques;
> Educational Programs in High Schools*

Summary 139

Review and Discussion 141

9 ARMED ROBBERY 142

Introduction 142

> *Basic Differences Between Robbery and Burglary;
> Business Robbery;
> Armed Robbery Is Not Always a Logical Crime*

Before the Robbery 145

> *Recognizing That the Business Is Being "Cased";
> Some Precautions Beforehand;*

Kidnapping, Coupled with Robbery;
A Reward Program Is Usually Ineffective

Minimizing the Loot   149

Setting a Cash Limit; Bleeding Cash Registers;
Protection Should Also Be Given to Checks;
Location of the Company Money Room;
The Holdup at Opening Time; Paying by Check

What May Be Done During the Robbery   152

Tripping the Silent Alarm;
Call Police Authorities Not the Company Auditor;
Giving a Signal to Co-Workers That a Robbery Is in Progress;
The Need for an Accurate
and Complete Description of Each Bandit;
Unusual Features of Bandit May Be Significant;
Observing the Bandit's Activities and the Objects He Touches;
Giving Out the Bait Money

After the Robbery   158

Recording Names and Addresses of Witnesses;
Advising the Police That the Robber Has Departed;
Interviews and Statements by Witnesses or Management;
Statement to the Press About the Amount of Loss

Summary   160

Review and Discussion   162

10 BUSINESS AND RESIDENTIAL BURGLARY   163

Background   163

What is Burglary?;
Conditions Behind Increases in Burglary;
Three Basic Types of Burglars;
What Are Real Losses in Business Burglary?

Three Basic Approaches to Prevent Burglary   166

Limiting the Time Available to the Burglar;
Limiting the Amount of Money or Loot That Is Available;
Hardening the Target;
Laws Requiring Hardening of the Target:
Leadership of Oakland, California, Police;

*Protecting the Safe Combination and Building Keys;
Some Techniques Used by Professional Burglars*

Solving Burglaries   170
Helping the Prosecutor Prove the Case

Summary   171

Review and Discussion   171

11   MERCHANDISE HANDLING—
RECEIVING, SHIPPING, WAREHOUSING   173

Dock Problems   173

Receiving   175
*Receiving "Blind" as a Good System;
Security Department Verification of Receiving;
Receiving Overships*

Shipping
*Shipping by Mail or by United Parcel Service (UPS);
Thefts from the Delivery Truck; Theft of the Entire Load;
Preloading of Delivery Vehicles;
Surveillance (Tailing) of Company Trucks;
Using Railroad Seals to Control Shipments*

Warehousing   183
*Selecting and Checking Shipments;
Controls over Returned Merchandise*

Summary   184

Review and Discussion   186

12   PERSONNEL SECURITY   188

Obtaining Honest Employees   188
*Interviewing Carefully; Investigations of Job Applicants;
The Polygraph; Psychological Stress Evaluator (PSE)*

Keeping Employees Honest   192
*Motivating Employees for Continued Honesty;
Letting Employees Know About the Security Program;
Reminding Employees of Security Responsibilities
in Awareness Sessions; Company Security Rules*

Dealing with Problem Employees   196
> *Helping the Problem Employee; Discharging the Thief;*
> *Temporary Employees;*
> *Some Examples of Problems Caused*
> *by Temporary Employees; Exit Interviews; Bonding*

Summary   201

Review and Discussion   203

13 COMPUTER SECURITY PROBLEMS   204

Background   204

Location of the Computer Installation
> *Physical Catastrophes; Physical Controls*

Fire as a Threat to the Computer   207
> *Location and Use of Detectors;*
> *Extinguishing Agents for Computer Fires;*
> *Water Sprinkler Systems; Carbon Dioxide Gas (CO_2);*
> *Halon 1301 (Freon); Hand Extinguishers and Floor Pullers*

Backup Records   214
> *Off-Site Storage of Backup Records; Backup Systems*

Further Preventives   215
> *Access to the Computer; Tape Library Controls;*
> *Disposal of Computer Printout Materials;*
> *Emergency Shutdown and Recovery Procedures*

Summary   216

Review and Discussion   218

14 PROTECTING TRADE SECRETS AND CONFIDENTIAL INFORMATION   219

The Security Threat of Business Espionage   219
> *The Nature of Industrial Espionage;*
> *What Is Business Espionage?;*
> *The Protection Philosophy of the Courts;*
> *Business Espionage Is Not New;*
> *Two Basic Security Problems;*
> *What Is Usually Taken in Business Espionage;*
> *Legitimate Sources of Competitive Information;*

*How Secret Information Is Developed; Document Protection;
Protecting Documents That Are Being Used;
Telephone Taps; Some Additional Techniques
to Prevent Business Espionage;
The Conflict Between Company Interest
and Employee Information;
Management Programs to Alert Employees
to Responsibilities;
Exit Interviews for Employees Working
with Confidential Secrets*

Emergency Readiness and Disaster Programs   228

Telephone Calls, Bomb Threats, and Bomb Problems   231

*Specific Instructions to the Telephone Switchboard;
Evacuation; Keeping the Bomber Out of the Building;
Reporting the Location of a Bomb*

Summary   233

Review and Discussion   235

15 OFFICE SECURITY PROBLEMS   236

Bad Checks   236

Protecting Corporate Valuables   239

Protecting Office Art and Decorations   240

Office Supplies   241

Petty Cash Funds   242

Mailroom Problems   243

*Using the Mailroom to Ship Stolen Items;
Incoming Cash and Checks in the Mailroom*

Business Machines   245

Payroll Frauds   246

Telephone Misuse   246

Summary   247

Review and Discussion   248

16 SPECIAL SECURITY PROBLEMS OF INSTITUTIONS 249

Background 249

Hospitals 250
> *Parking and Escort Problems; Access Controls; Theft Problems; Package Passes; Thefts from Patients; Payment for Hospital Medication and Services; Disaster and Evacuation Plans*

School Security 254
> *Additional Problems in College Campuses*

Museum, Library, and Park Security 258

Summary 258

Review and Discussion 260

17 THE OCCUPATIONAL SAFETY AND HEALTH ACT (OSHA) 261

Purpose and Broad Coverage of the Law 262

Responsibility of Employers under OSHA 263

Administration of OSHA 263

OSHA Violations 264

Time for Correction of Hazards 265

Record-Keeping Requirements 266

The Extensive Number of OSHA Standards 267

Summary 268

Review and Discussion 269

18 SECURITY MANAGEMENT AND ADMINISTRATION 271

Introduction 271

The Security Director's Job 272
> *Enlisting the Cooperation of Other Departments; Organizing the Security Department;*

*The Loss Prevention Approach;
Justifying the Security Program;
Managing Undercover Operations Within the Company*

Private Security Problems 277

Differences Between Sworn Police Officers and Private Security; Career Opportunities in Security

Summary 279

Review and Discussion 281

APPENDIX 1 SPECIFICATIONS FOR CHAIN-LINK FENCING 282

Fence Gates 283

Signs on the Fence 284

Fence Alarms 284

APPENDIX 2 MISCELLANEOUS FORMS 285

Salvage or Scrap Material Sales Form 285

Description Form to Be Filled Out by Witness 286

Telephone Operator's Bomb Threat Report 287

Statement of Business Security 287

APPENDIX 3 MODEL BURGLARY SECURITY CODE 290

Preface

It is the intent of this book to provide a general background in security. Written for students of Criminal Justice and Police Science, it was also intended for individuals desiring to work in the private security sector. In addition, it was designed for students in management and business careers, to understand how security functions in the business environment.

It is also hoped that this work will have practical application for security administrators, architects, engineers and others with responsibility for the protection of industrial and business buildings and organizations.

Sharp increases in personal and business crimes have emphasized the need for cooperative action between public police agencies, business officials and the private sector of security. There has never been a greater need for protection. How this may be accomplished is the approach that this book seeks to develop.

I wish to thank Mr. Russ Hawe, Adams Rite Manufacturing Co., Professor Derald D. Hunt of Golden West College, and Mr. Jack R. Brick, Director of Public Safety, University of California, Long Beach, for suggestions and helpful ideas. Margaret A. Morse afforded considerable assistance in manuscript preparation, and I am indebted to Anita M. Hemphill and Phyllis Hemphill for editing.

CHARLES F. HEMPHILL, JR.
Long Beach, California

Humanity's Long Struggle to Maintain Security

People have always had to guard their lives and material things. For thousands of years, tree dwellers built their homes above ground to keep out enemies. Early Europeans built shelters on piles, supporting platforms off the shore of a lake. Access could be had only by drawbridge or boat. Early cliff dwellers in the United States built homes and food storage bins in great crevices high in the walls of cliffs. These locations could be reached only by ladders that were drawn up into the structures by the occupants.

Two thousand years ago, the Chinese built a great wall to keep out marauding Mongol tribes. This wall was approximately 25 feet high and 1500 miles long. The building of this structure required the services of half a million workers for 15 years.

In ancient times almost all towns and cities were surrounded by walls that were guarded almost continually. Well into the Middle Ages, feudal castles continued to provide security with thick stone walls, surrounded by deep moats filled with water.

In comparatively recent times, frontier outposts in the United States used wooden stockades with block houses and rifle ports at the corners. Today, this security aspect has carried over into the protection of individual homes, businesses, and institutions.

Two Basic Kinds of Security

Security in the United States may be divided into two broad classes, depending on the source from which the service originates:

1. Police agencies, prosecutors, and the courts combine in a common goal—to preserve the peace and to keep all areas of the community safe. Because this part of security results from the efforts of public or governmental officials, it may be referred to as the legal, or public, sector of security.
2. The other main area of security is that furnished by private security companies or security departments of individual companies. This is usually called the private sector of security.

A generation or two ago, many individuals regarded all security as a police problem, rather than one to be solved by both governmental agencies and private security representatives. Today, it is realized that

1

Introduction

BACKGROUND AND GENERAL PROBLEMS OF BUSINESS AND INDUSTRIAL SECURITY

A Definition of Security and What It Implies

"Security" is defined as protection; assurance;[1] or a state or sense of safety or certainty.[2] The word "secure" means not exposed to damage.[3] Security implies a stable, relatively unchanged atmosphere in which individuals or groups may pursue their ends without disruption or harm, and without fear of loss or injury.

No one can provide absolute protection for a business or industry, an institution, or a home. But an environment can be maintained in which loss or harm is not likely to occur. This is security.

---

[1] *Black's Law Dictionary*, 4th ed., s.v. "security."
[2] *New World College Dictionary*, 2nd ed., s.v. "security."
[3] *Webster's New World Dictionary of the American Language, Second College Edition*, s.v. "secure."

there is a pressing demand for both kinds of security, with the contribution of one class supplementing the work of the other.

The need for protection has grown to such proportions that it is simply beyond the capabilities of police forces alone. In recent years, police attention has shifted from the prevention of crime to responding to crimes already perpetrated. In protecting their investment, private firms have turned to the private security industry.

The Increase in Crime—
Its Impact on Business

No one knows the real cost of crime in this country. Business sources can only maintain lists of known losses and submit estimates. U.S. Department of Commerce statistics for 1973 showed losses of $18.3 billion dollars,[4] which was a staggering increase of 31 percent over 1971 totals.

Moreover, these Department of Commerce statistics, based on law enforcement agency reports, are believed to be conservative estimates since many crimes go unreported and are not included in these figures.

In 1975 the Small Business Administration reported that criminal losses cause 30 to 40 percent of all small business failures.

Companies that manufacture, distribute, or sell merchandise to the public are particularly open to loss from their own personnel, delivery people, customers, and professional criminals. Owners and managers are often caught up in the day to day problems of operating their own firms. They are usually aware of the likelihood of loss, but they do not understand how losses occur and what can be done about them. Then, too, there are firms that do not deal directly with the public, but that occupy ever larger buildings, housing employees who must be protected against theft, criminal assault, fire, and other catastrophes.

Tax dollars simply will not stretch far enough to provide police protection for all of the security needs of business. In addition, private security operates in many areas where regular police officers cannot function, either legally or from a practical standpoint. A private security guard, for example, can enforce company rules under the direction of a business or industrial employer. Police officers have no right to enforce business rules that are not backed up by the force of law, regardless of management's desires. In many situations, a policeman simply has no authority to go on private property, or to take action after entering.

Then, too, it is no longer adequate merely to catch criminals or dishonest employees. The security function of today must contribute to

---

[4]More recent statistics are not available.

profitability. Business and industrial establishments need kinds of protection available only from private security forces. The public, or police sector of security, must work closely with the private sector, since neither class alone can fill all the needs of society.

HIGH POINTS IN THE HISTORY AND DEVELOPMENT OF PRIVATE SECURITY

In the first century A.D., the Roman Emperor Caesar Augustus hired nightwatchmen to protect businesses and homes in his capital city of Rome. Little is known as to how these watchmen functioned, but they seem to have been forerunners of both the private security industry and modern police agencies. Collapse of the Roman Empire brought this early activity to an end.

Organized security, as we know it, was apparently almost nonexistent from the time of the Romans until merchant guilds and trade organizations became established on the continent of Europe and in England. In the fourteenth century, guild members at locations in Germany and England banded together in local groups, hiring watchmen to protect their retail shops and business warehouses during night hours.

Toward the end of the eighteenth century, there occurred in England a series of changes in methods of industrial techniques and economic organization that has been termed the Industrial Revolution. These transformations also brought sudden changes in many other aspects of life. People from the countryside began to flock to the cities, and London, Birmingham, and the other English population centers were simply not able to take care of the sudden increase of residents. Social conditions quickly became disorganized. Living quarters were poor and overcrowded. Sewage disposal and sanitation were serious problems. Public transportation between housing and job locations was almost unknown, and getting food and necessities into the populated areas was slow. Working long hours under unsafe conditions, the laborers often fared poorly. Social conditions became disorganized.

These unsettled conditions contributed to a sudden increase in crime throughout industrial areas in England. And with crime on the rise, tradesmen and merchants were most often singled out by thieves, burglars, and armed robbers. The need for the development and expansion of a private security industry was apparent.

In 1737 the British Parliament provided some governmental help by allowing public tax monies to be used for the first time in hiring of nightwatchmen for public streets. Eight years later, Parliament approved a plan to combine the resources of several newly organized private security forces in setting up a protection system for businesses. These private security groups failed to work together effectively, however, and little came from this early day plan.

Colonists coming to the New World banded together for mutual safety and business protection. Local governments in the American colonies formalized a watch system, requiring every male inhabitant to serve regular turns as a nightwatchman. The tour of duty usually began at 9 or 10 P.M. at night and ended at sunrise. During an assignment of this kind the watchmen sometimes came face to face with burglars, thieves, marauding Indians, wild animals, grave robbers, or runaway slaves. The watchmen also put out fires, or spread the alarm if the blaze was discovered too late to control.

By 1800, merchants in New York, Boston, Philadelphia, and other trade centers hired watchmen on a monthly salary, or maintained family living quarters on the floor directly above their street-level shops.

In 1850, Allan Pinkerton founded a private business for security in Chicago, offering to do more than furnish nightwatchmen. Pinkerton agreed to handle business investigations when needed by his clients, and he obtained the business of several midwestern railroads and industrial firms.

When the Civil War broke out, George B. McClellan, a former vice president of the Illinois Central Railroad, was commissioned a major general in the Union Army. Accustomed to using Pinkerton's *investigators* for railroad inquiries, McClellan hired Pinkerton's agents to gather intelligence information for the Northern Army. The Union forces had no spy system prior to that time.

In 1858, Edwin Holmes began the operation of the first central station burglar alarm operation. This eventually evolved into Holmes Protection, Inc.

In 1859, Washington Perry Brink organized a package delivery service in Chicago. This served as the forerunner of modern armored car services when Brink delivered his first payroll in 1891.

As business developed after the Civil War, some industrial firms hired their own company security guards. Used as an armed strikebreaking force in 1892, company guards at the Homestead Steel Company, Homestead, Pennsylvania, shot and beat striking steel workers. Similar attacks were made against striking workers by company security forces in the troubled automobile industry during the middle 1930s.

Throughout the history of this country, private security firms have consistently refused to participate in organized lawlessness. The Homestead and Detroit incidents have been remembered as unlawful excesses that in no way represent basic attitudes in the security industry.

In the decade before 1900, criminals throughout the country began to attack rail shipments with increasing regularity. As theft losses for unaccompanied freight shipments mounted, many state legislatures passed statutes known as the Railway Police Acts. These laws enabled the railroads to organize and control their own private police forces that had been granted law enforcement powers.

This was a logical outgrowth of the problem, since most of the rail lines ran through so many different jurisdictions that security would have been difficult to maintain without a unified force. By 1921, the Association of American Railroads furnished coordinated security services to all principal rail lines in the country.

The outbreak of World War II, December 7, 1941, gave added impetus to the development of private security in the United States. German undercover agents in America had committed numerous acts of sabotage during World War I, and most of these incidents had not been recognized as sabotage at the time. In the interval between the two World Wars, considerable publicity had been given to these World War I attacks. By 1941, British and American intelligence sources were in possession of considerable evidence that Nazi Germany intended to wage a systematic campaign of sabotage against American warehouses, shipping facilities, and factories producing goods for the war effort.

Being aware of these Nazi intentions, the U.S. government required many factories to set up employee identification and access systems, to conduct plant protection surveys, and to utilize uniformed guards. Individuals who were not acceptable for military duty were pressed into service, guarding strategic railway bridges, tunnels, airfields, and dock areas.

As needs continued to expand, new security firms sprang up all over the country following World War II. Many of these new companies filled business requirements but were hampered by lack of capital, management experience, and trained personnel. A number of these new firms were incorporated into security companies with more resources, becoming today's leaders in the security industry. As an example of the trend toward acquisition, Baker Industries purchased Wells Fargo. This allowed Baker Industries to offer customer service beyond its electronic detection and equipment origins, providing guards, armored car service, patrol, and investigative capabilities.

In 1955, the American Society for Industrial Security (ASIS) was founded, with headquarters in Washington, D. C. Organized to "foster and enhance the highest degree of professionalism in all fields of security," the society was influential from the outset in raising standards, in educating, and in obtaining deserved recognition for the security industry. ASIS has contributed much in providing professionalism and in gaining the respect of business.

During the 1950s and 1960s, insurance companies began distributing materials of considerable value to business, highlighting methods of control that could be used by security or business people seeking to reduce losses. A helpful periodical of this kind is typified by "Ports of the World," published by Insurance Company of North America (INA). This publication is a guide to cargo loss control, with an analysis of shipping losses attributable to theft and other crimes against business.

Race riots and student disturbances in the 1960s and early 1970s highlighted additional needs for private security. Beginning with arson, looting, and indiscriminate shootings in the Watts section of Los Angeles in 1965, racial disturbances quickly spread to Detroit, Newark, Washington, D. C., and other parts of the country. A number of lives were lost in these riots, and large areas of commercial and residential property were destroyed.

On February 11, 1969, a Control Data Corporation 3300 computer was destroyed by rioting students at Sir George Williams University in Montreal, Canada. In the same year, five members of an antiwar group in the United States, calling themselves "Beaver 55," broke into the Dow Chemical Corporation's data research center at Midland, Michigan, erasing or destroying scientific data on approximately 1000 reels of computer tape. A number of other attacks followed, in which U. S. students smashed computers, threw acid into printed circuits, used explosives, and attacked research and computer installations with fire bombs. Explosive charges were set in a number of college and university buildings, as well as structures owned by major corporations.

Kidnappings by terrorist groups became quite common in some parts of the world in 1971. Incidents of this kind occurred frequently in Argentina, Uruguay, and in other countries where life and death political struggles were taking place. Terrorist kidnappings of wealthy industrialists in Mexico brought this crime close to the United States, while kidnappings for ransom continued throughout this country.

In 1968, radical groups had also begun the hijacking of airplanes in a number of different countries, seeking to force their demands on the rest of the world. By the early 1970s, commercial airline hijacking was out of control. On a trial basis, the Federal Aviation Agency set up a passenger predeparture screening program at Honolulu, Hawaii, in June, 1972. Results indicated that commercial passengers could be screened with a minimum of inconvenience and that the process was effective in spotting passengers with guns, explosives, or other weapons. On the basis of the Honolulu tests, the Federal Aviation Agency ordered the screening of all commercial U.S. airports on January 6, 1973, The effectiveness of this national security program became evident almost immediately.

THE SERVICES PROVIDED BY PRIVATE SECURITY

A wide variety of services are regularly furnished by private security firms to businesses, industrial establishments, institutions, and individuals. Only a partial list of those relying on protective services of this nature would include retail stores and trade associations, wholesale establishments, docks and shipping facilities, trucks, planes and freight vessels, airports, educational and public institutions, hospitals, museums, hotels and motels, offices and high-rise buildings, theaters and sports stadiums, construction sites, places of public and private amusement, libraries, homes, apartments, and condominiums.

Some private security companies provide only a limited number of security services, whereas other firms have widely diversified programs for all types of clients. Typical services provided are described below.

Guard Services

Guard supply is the type of security service most frequently furnished to business or industry. Practically all guards are uniformed, since the wearing of the uniform adds authority to the guard's responsibility.

A guard furnishes company employees with a sense of security, and gives a headstart in reporting and controlling fires. Security representatives of this kind are also particularly valuable in discouraging and preventing payroll robbery, nighttime burglary, and other serious crimes. Properly supervised and directed, guards may also reduce the incidence of internal pilferage and theft.

Used in other functions, a guard may control access, escort visitors within the facility, and regulate compliance with company rules. The guard may also be an ambassador of company good will, furnishing information, assisting visitors, and regulating foot traffic and vehicles. If performing public relations duties, the guard may be outfitted in a blazer and slacks rather than in a conventional uniform.

Patrol Service

A patrol service offers inspection tours of the property to be protected, usually with a specified number of tours to be made during the night hours, or within a 24-hour period. Patrolmen in cars or on motorcycles can give coverage to a large number of customers, although a foot patrolman may observe more detail and may not be so readily spotted by criminals loitering in the neighborhood.

Costing considerably less than a full-time guard, a patrol service cannot be expected to afford the same continuous protection as a guard. The difference here is that an alert criminal may remain hidden in shadows or shrubbery outside a business and can observe the rounds made by a patrol service. The criminal will then break into the building, hoping to leave before the patrol service makes another inspection tour.

Investigative Services

With the expansion of American business, management needs more information to assist in making decisions. A security investigator is often of considerable help in this search for facts.

A well-staffed security company may be equipped to conduct background investigations of:

1. Potential executives, to determine ability, character, experience, stability and personal habits.
2. Employees, to ascertain reliability, financial status, ability, potential for advancement, and character traits.
3. Outside businesses, to determine the management ethics, ability to pay bills, reliability, and past performance.
4. Organizations and associations, to ascertain the identity of officers, financial supporters, objectives, community status, purposes and methods of operation, and past conduct.

A capable security investigator may also be called on to obtain information about stockholders' disputes and proxy matters, to investigate jury panel members for lawyers, or to ascertain the financial responsibility of the defendant in a lawsuit.

At other times, a private investigator may conduct surveillances to develop facts about persons involved in insurance claims, in child custody and domestic matters, in questionable employee activities; and in occasional verification of the activities of deliverymen and truck drivers.

In addition, the investigator may make surveys to determine product acceptance, surveys as to the honesty, courtesy, efficiency, and appearance of employees, or public opinion surveys. An investigation may also be made of the amount of traffic passing in front of a location contemplated as the site of a new business.

This is only a partial list of the kinds of inquiries that may be handled. The types of investigation may be limited only by the ingenuity and experience of the investigator, as well as by the needs of business.

Armored Car Service

An efficient, reputable armored car service can furnish effective protection for any business that accumulates or pays out unusual amounts of money. Wells Fargo Armored Car Services, for example, point out that their armored vehicle is a "vault on wheels" at the door of the businessman. This firm offers a money or valuables depository service, pick up and delivery of currency or valuables, and courier service to accompany valuables in transit. In addition, Wells Fargo Armored Car Services points out that they handle payroll delivery, check cashing facilities, coin collection services, money sorting, counting, and packaging. In addition, this firm advertises that bank-type vaults are provided when funds or valuables are to be retained overnight or longer, as well as two-compartment safes that may be consigned to a customer.

The Central Alarm Station

In some communities an alarm system may operate on a direct line to a police department switchboard. With increasing demands on police facilities and personnel, this direct notification to a law enforcement agency is being discontinued in many areas of the country.

A central alarm station, operated by a private firm, maintains a central alarm board, monitoring signals set off by sensors on the customer's property. These signals are transmitted to the central station by leased telephone line. Some central alarm station companies do not dispatch their own personnel in response to an alarm but instead relay the information received to the appropriate police agency for action. Most security companies that operate a central station, however, dispatch their own uniformed runner or patrol, in addition to calling the police for a response by sworn officers.

While most alarms transmitted to a central station are silent signals designed to notify of the presence of intruders or burglars, the central station may also be used to receive signals from fire sensors on the client's premises, as well as to monitor water pressure lines, temperature gauges, or almost any critical industrial process that should be regularly verified or monitored.

Undercover Operators

Many security companies can provide undercover operators to work in the client's business, according to management needs. Often an operator of this kind can find answers to questions that management does not want to ask directly.

The most obvious information developed by an undercover agent usually concerns employee theft, dishonesty, or drug addiction. A competent operator may develop other kinds of information that are also of value to management. Information may be obtained about employee morale, production slowdowns, falsification of production or shipping records, or details about improper receiving and shipping. In addition, the undercover employee may be able to learn why some management projects fail while other programs are accepted by employees and are successful.

Loss Prevention Surveys

A limited number of private security firms and consultants have the capability to conduct loss prevention surveys for business or industrial establishments. The focus here is on prevention rather than investigation after a specific loss has occurred. Working closely with management, this type of professional determines where losses are occurring; the consultant then devises methods to control these losses. Success here depends on a wide knowledge of both security and business management. It is also necessary for the consultant to get both management and employees to institute and use the controls that are necessary to bring about changes in employee actions and attitudes.

Predeparture Screening

In the five years prior to 1973, there were 147 attempts to hijack U. S. airplanes, and 91 of these attempts were successful to some extent. Since January, 1973, private security firms have conducted predeparture screening operations of all department passengers and their personal belongings. Operating on private contracts with individual commercial airlines, these security companies have been highly effective in making certain that no weapons, explosives, or potentially dangerous devices or substances are taken aboard airplanes. Private security firms have also been successful in reducing baggage thefts at commercial airlines.

Guard Dogs

Guard dogs are frequently made available by security firms, either with or without a handler to control the animal. Locked inside a yard or building, the animals are usually very effective in keeping intruders out. However, dogs may frequently endanger janitorial or maintenance employees, and a serious lawsuit may result if an animal escapes or attacks someone who has a right to be on the premises.

Although trained dogs have proved very effective in searching a building or warehouse where a criminal may be hiding, hiring such dogs and their handlers is often too costly to justify the practice for most guard or patrol work.

Polygraph Examinations (Lie Detector Tests)

Security firms in many areas have licensed, trained polygraph operators, who can provide a variety of examinations. The results of the polygraph (lie detector) test are not usually admissible as evidence in a court of law. And in some localities a polygraph examination cannot be required of an applicant for a job.

If applicants take such a test voluntarily, however, the test results can help the employer in selecting loyal, honest, and competent employees. The applicants' intentions as to job permanency can also be ascertained. In addition, specific examinations to determine guilt or innocence of employees suspected of dishonest acts can be helpful.

Executive Protection

A limited number of security firms are able to provide comprehensive programs for businessmen and family members who may be subject to kidnapping, assault, or blackmail attempts.

Protection of individuals, automobiles, and buildings against bombing attacks and arson is also a specialized class of security service.

Shopping Services

Shopping services utilize crews that test the honesty of sales clerks in retail stores, bars, restaurants, and other sales establishments. These crews also audit and report on the efficiency, courtesy, and attitudes of sales or service personnel.

In testing honesty, shopping crews usually simulate actual purchase transactions, giving the sales clerk an opportunity to steal by failing to ring up cash on the cash register, by underringing, or by other dishonest manipulation. The shoppers who make these tests are dressed as customers, giving no indication that the transaction is not completely normal.

Retail stores or bars usually contract for services of this kind on an annual basis, with a specified number of tests in a particular store or bar. If dishonesty is detected, or an apparent failure to ring up receipts from a sale is observed, a different shopping crew is normally utilized to make certain the first crew was not in error.

Shopping services are usually operated by security firms that specialize in this kind of service to merchants.

"Sweeps" for Bugging Devices

A limited number of security firms also have the ability to make "sweeps" to determine whether hidden microphones or listening devices have been installed in company board rooms or other critical locations. The experts employed by these firms are also able to determine whether a client's telephone has been tapped.

Safety

Some private security firms are prepared to set up a company first-aid department, to furnish ambulance service and trained first-aid crews, and to provide safety surveys and educational training for a client's employees in this field.

Fire Protection

A limited number of security companies can provide personnel well trained in the technical field of space-age fuels, or a complete municipal or industrial fire department, staffed and trained. In addition, experts can be made available to perform fire prevention surveys and offer educational programs in this field.

Psychological Stress Evaluator (PSE)

Some security firms can provide psychological stress evaluation tests of individuals suspected of theft or misconduct. The theory behind these tests is that stress, induced by fear, facilitates detection of attempted deception during interrogation. As with the polygraph test, an examination of this kind should not be given by anyone other than a trained operator.

This list of security services is not all inclusive. Other types of services and investigations can be made available. Scientific developments will further expand the types of security services that may be available to clients in the future.

In recent years, women have served alongside men in all types of security jobs, and in some companies, women have been employed since about 1940. Their assignments have not been restricted to clerical functions or responsibilities such as predeparture screening in commercial airports but have included every type of security job for which the individual was qualified. Women security guards are sometimes especially useful in controlling emotionally charged men who threaten immediate violence.

By the end of 1976, the U. S. Department of Labor reported that approximately 800,000 individuals, young and old, male and female, were employed in private security throughout the United States. In all likelihood, the demand for trained individuals in the security field will increase greatly in the immediate future. Additional information as to career opportunities in security is set out in other sections of this book.

SUMMARY

The word "secure" means not exposed to damage. Security implies a stable environment in which goals may be pursued without disruption or danger, without fear of loss or injury. Absolute protection for a business, industry, or home does not exist. But an atmosphere can be maintained in which loss or harm is not likely to occur. This is what we mean by security.

There are two basic kinds of security: (1) that provided by police or public agencies and (2) that furnished by private agencies. There is a real need for both public and private security, with one supplementing the work of the other.

The impact of crime has been enormous in the business community. Criminal losses cause 30 to 40 percent of all small business failures, according to the Small Business Administration. This pressure from the criminal element intensifies the need for private security.

Throughout the history of private security, racial riots, student attacks, wartime sabotage, terrorist hijackings of airplanes, and executive kidnappings have all contributed to an increase in the need for private security.

Private security companies offer many services to businesses, individuals, and the owners of private properties. Included in these services, among others, are uniformed guards, patrol services, guard dogs, central alarm station services, investigative services, undercover operators, loss prevention surveys, predeparture screening for airlines, armored car delivery service, polygraph (lie detector) tests, executive protection surveys, shopping services, and "sweeps" for hidden microphones.

By the end of 1976 there were approximately 800,000 individuals employed in private security firms, with an increasing demand for more services.

REVIEW AND DISCUSSION

1. Define security in your own words.
2. What are the two basic kinds of security? Describe the differences in the two kinds.
3. Is absolute security ever obtainable, or is it relative? Why?
4. List some of the high points in the development of private security as an industry in America.
5. List ten typical kinds of security services that are made available to clients by private security firms.
6. What is the basic difference between a guard service and a patrol service?
7. Describe how a central station alarm system works.
8. What benefits can management expect from an undercover operator in a business?
9. Give a brief description of the benefits that can be derived from a shopping service.
10. Explain why computer security surveys may increase in importance in the future.

GOOD GENERAL SOURCES ON SECURITY

Richard J. Healy, *Design for Security* (New York: Wiley, 1968).

Gion Green and Raymond C. Farber, *Introduction to Security* (Los Angeles: Security World Publishing Co., 1975).

Richard B. Cole, *The Application of Security Systems and Hardware* (Springfield, Ill.: Charles C. Thomas, 1970).

Charles F. Hemphill, Jr., *Security for Business and Industry* (Homewood, Ill.: Dow Jones—Irwin, 1971).

Timothy J. Walsh and Richard J. Healy, *Protection of Assets Manual* (Santa Monica, Calif.: The Merit Company, 1974).

Mary Margaret Hughes, ed., *Successful Retail Security* (Los Angeles: Security World Publishing Co., 1974).

Charles F. Hemphill, Jr., and Thomas Hemphill, *The Secure Company* (Homewood, Ill.: Dow Jones—Irwin, 1975).

2

Physical Facilities

BUILDING LOCATION AND CONTROLS

There are some benefits obtainable from physical security, but there are also problems that may be involved. The site and its surroundings almost always influence requirements for protection. We consider building placement or location, as well as building usage on the property, here. We also examine physical devices and controls that may be used to regulate entry into the location. Installation, design, and proper usage of doors, windows, and other openings in the building shell may contribute substantially to an effective protection program.

Physical Controls—Deterrents to Entry

Physical controls alone are only part of a good protection program. Ordinary barriers will not stop a skilled burglar equipped with modern tools, provided the burglar has time to use the tools. But good obstacles to entry make it so difficult that intruders do not want to make the effort to get in or risk getting caught. In face of a determined attack, however, physical security should be relied on only to deter or impede.

Three Lines of Defense for Physical Security

Authorities on security generally point out that there are three lines of defense where physical defenses or protective features may be utilized to provide security. These are

1. Perimeter barriers, usually located at the outer edge of the property.
2. Exterior walls, ceilings, and floors of buildings that are to be protected.
3. Areas within the building itself that can be utilized to provide security.

Choosing the Neighborhood and Planning Building Placement

Site selection should be carefully considered in choosing a location for a business, an institution, or a private home. By furnishing specific figures on crime rates in a given neighborhood, most city governments can usually provide practical guidelines to location. Information as to police patrols and coverage in the area may also be helpful in making a site selection. If employees, especially women, are required to work night shifts in a crime-ridden neighborhood, the employer may find it difficult to hire and retain satisfactory workers.

Placement of the building on the property itself is also worth study. Each structure should be positioned on the lot to afford a maximum view to each passing police patrol vehicle.

It is also suggested that the architect eliminate hidden recesses, abutments, projections, or other building features that would allow concealment of intruders. Experience shows that in most instances the two police officers in one patrol car can adequately cover the exterior of a building from opposite corners. But the effectiveness of this coverage can be canceled out by architectural features that allow an outsider to hide on the premises. If building walls are painted a light color, it is usually easier to spot the outline of an intruder on the premises at night.

Enclosed, lighted walkways between buildings are a way of providing security in a business or industrial complex.

Landscaping

Landscaping around a building is, of course, a definite attraction. But masses of shrubbery or trees growing near a structure can provide cover that would allow a criminal to attack a building wall. Shrubbery or any other visual obstruction will make doors and windows more vulnerable. It is therefore recommended that plants and trees should be no closer than 40 feet from a building, with ground foliage and shrubbery limited to approximately 2 feet in height.

A tree or telephone pole located near a building provides easy access to a building roof, an area of the building that can be cut through with comparative ease.

Wooden pallets, boxes, skids, and similar items may also cause problems if stacked alongside a building. These accumulations may be climbed to the roof, or may give cover to someone pounding a hole in the building wall. It should also be pointed out that an arsonist or a disgruntled employee may set fire to stacks of wooden pallets or boxes, intending for the fire to spread to the building.

Planning the Building Proper for Good Security

In planning a new business facility, the needs of customers and employees are a prime consideration. Efficiency and convenience are taken into account, along with accommodation of the work to be accomplished. But it is not unusual for security planning to be ignored until the facility is nearing completion. Sometimes this happens because business officials assume that security requirements will be automatically considered by the architect, along with other problems of construction. But even when adequate security features are planned for the building from the outset, they are often the first items to be struck out if the construction costs begin to exceed estimates.

Frequent contact with the building's architect is to be recommended, beginning with the consideration of the overall site. The architect should be made aware of all aspects of the problems involved in determining building placement, perimeter controls, fencing, lighting, parking areas, entrance controls, and the degree of protection to be given to sensitive areas.

Planned protection pays. And protection must, of course, anticipate future changes and developments in criminal technology.

PERIMETER CONTROLS

We have seen that perimeter security keeps unwanted persons from coming on the property but that this protection alone may not automatically result in good security. There is still the likelihood of internal losses from employee theft or other inside causes. But in most situations, perimeter security is worth the effort and cost.

Many kinds of protection barriers may be used: rivers, cliffs, canyons, dense stands of foliage, or combinations of such natural features.

Frequently, however, a retail store or large office building will cover all of the real estate on which it is located. In a situation of this kind, the outer walls of the building itself may be considered as the perimeter of the property.

In the material that follows, we discuss the fence, the most commonly used type of perimeter barrier. Then we examine the security problems involved in protecting doors, windows, and other openings in the outer shell or building ceilings and walls that can be regulated by exit and entry controls and physical devices.

The Fence as Perimeter Protection

A fence may be used both to prohibit unauthorized entry and to regulate travel of those with a right to be on the premises. It is the first line of defense in protecting shipping areas, parked trucks, and employee vehicles.

Chain-link fencing is frequently used for security fencing purposes, since many security authorities feel that chain-link fencing is superior to a masonry wall or a wooden fence. The most important advantage is that chain-link fabric allows passing police cars, company guards, or any employee to see what is happening on both sides of the perimeter barrier.

A masonry wall or wooden fence does not, of course, offer this easy visibility, and other types of woven wire fencing may not be as strong as chain-link. It is worth noting that while some masonry walls and wooden fences are quite strong, they also provide a steplike arrangement that can be used to climb over the barrier.

The visibility offered by a chain-link fence will, of course, be compromised by shrubs, vines, or trees allowed to grow into and alongside the fence. Decorative plastic inserts in the fence fabric will also hinder visibility.

Often business officials object to chain-link fencing because it creates an "institutionlike atmosphere" or because it fails to hide the industrial aspect of a business. Where appearance of a business or a facility is a major consideration, there are many materials other than chain-link fencing that may be used. Some decorative steel panels, for example, are both attractive and useful from the security standpoint. But they may not allow for as much visibility as chain-link fencing.

Recommended specifications for chain-link fences are set out in Appendix 1.

BUILDING OPENINGS

In general, the physical security of a building may be no better than the protection afforded to doors, windows, or other openings in the building shell.

Burglars and thieves break into buildings in many ways, but they seldom attack the walls of the structure. This does not mean that criminals will never pound through the sides of a well-built building, or that they will not tunnel under the floor. If a business shares a common wall with another store or business, there is an increased possibility that the wall may be attacked during the weekend or nighttime hours by someone who has rented the adjoining property.

In most instances, however, the intruder looks for easier ways to gain entry. The roof, for example, is usually easier to penetrate than the walls. If the building is one in a row of structures, access to the roof may be comparatively easy from an adjoining building. Here, a chain-link fence around the roof perimeter may make access difficult from a nearby structure.

Once a would-be intruder has reached the roof, there is little possibility of detection. Usually, the building's protective lighting is located below the roof line, and the lights of a patrolling police vehicle are effective in illuminating ground-level areas only.

Perhaps the majority of business alarm systems are perimeter alarms, making use of contacts on windows and doors. Entering through the roof will not break the contacts on the door or windows, and therefore the alarm will not be set off.

Once on the building roof, a burglar can quickly cut through a typical wood and asphalt roof, using only a carpenter's brace, a wood bit, and a small keyhole saw. The usual technique is to bore a string of holes about 1 inch in diameter, approximately 6 inches from center to center. These holes are bored in a square pattern, approximately the size of the desired entryway. The would-be intruder then uses the keyhole saw to cut from one drill hole to the next. When the entire square has been sawed away, the intruder will drop down through the hole into the interior of the building. If the roof is unusually high, a rope ladder may be used to climb downward.

It is through existing openings in the building shell that most unlawful entries are made. Doors, windows, skylights, vents, basement coal chutes, roof hatches, and the like are all vulnerable.

Available statistics maintained by insurance companies from 1950 to 1970 showed that existing openings were penetrated in about 90 percent of all break-ins. A survey of 2800 separate attacks against buildings in 1976 reflected that at least 30.7 percent of these entries were through doorways, 48.9 percent were through windows, and 2.9 percent were through roof hatches, skylights, vents, and transoms. Authorities on business and residential burglary are generally convinced that criminals will continue to make a high percentage of penetrations through existing openings.

From these figures, it is apparent that good security protection should be given to doors, windows, and other openings.

Building Exit and Entry Controls

From a physical security standpoint, a building should preferably be constructed of steel-reinforced cement, with floor, walls, and roof of the same material. Ideally, all windows should be eliminated and access restricted to a single door. Unfortunately, a building of this kind would not be very practical. It is possible, however, to limit the doors to the number needed for operational necessity. If possible, all doors except one should be locked from the inside. The remaining door should be located on a well-lighted, police-patrolled street and should be used by all employees for entry and exit. This single control door may then be equipped with an alarm system, time lock system, or other protective features.

When a building has few doors and they are well protected, the question may arise as to how the fire department could enter in the event of an emergency. Fire departments in almost all cities are now equipped with power saws that will cut through any hardened steel lock in a matter of seconds. As a practical matter, it is usually more economical to repair a break made by cutting through the bolt of a storefront or residential lock with the power saw than to repair the damage done by breaking in with axes, wrecking bars, or other heavy tools.

Fire exit doors as shown in Figure 1 are usually required by municipal fire and building codes for a specified number of square feet of floor space in any business or industrial building. The larger the building, of course, the greater the number of emergency exits that must be provided. While fire doors are necessary for safety, they are undesirable from a security standpoint. It is generally acceptable for these fire exits to be locked from the outside, but openable by pushing against a panic bar or crash bar on the inside. Under the fire codes in most cities, exit doors can be made lockable from the inside when employees leave the premises.

From a security standpoint, it is frequently desirable to connect an audible alarm to fire exit doors. Most of these alarms are so loud they can be heard by all employees and members of management for a considerable distance. Some can be shut off only by a supervisor or manager who has a key.

But merely alerting employees to the fact that someone has gone out an alarmed door is not sufficient. There must be some system to assure

22 Physical Facilities

FIGURE 1 An emergency exit alarm, openable from inside, but locked from the outside. Most devices of this type sound a loud alarm when someone goes out.

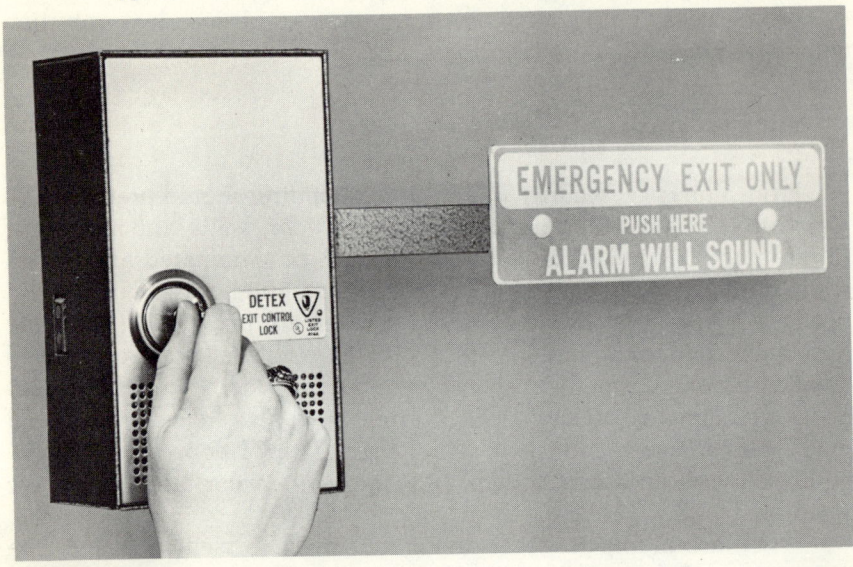

Photo courtesy of Detex Corp., Chicago, Illinois.

that a trained guard or supervisor immediately responds and determines who has breached and why. Experience shows that unless there is a response procedure, employees in a warehouse or business will use the alarmed door to carry out stolen merchandise.

If there are a large number of exit doors, it may be desirable to have each door wired to a central control panel (annunciator), flashing a light on a panel and sounding an audible alarm. A monitor panel of this kind, shown in Figure 2, will show the location of the door where the unauthorized exit was made.

Fire door exit alarms usually fall into one of three classes: (1) alarms activated by a spring windup, (2) alarms activated by a self-contained battery, or (3) alarms activated by standard electrical current, with a battery backup for emergency power failure. All three types perform well in practice, but it is advisable to make regular tests of the windup mechanism and power batteries that may be involved.

Padlocking All Doors Inside Except the Lone Control Door

We have already stressed the importance of padlocking business or warehouse doors from inside the building, where possible. This is so an outside criminal cannot manipulate or get at the lock.

FIGURE 2   Electrically wired monitor or annunciator panel, showing the location of the door where emergency exit alarm was breached.

Photo courtesy of Continental Instruments Corporation, Hicksville, New York.

In many installations a door is secured on the inside by means of a sliding bolt, or by inserting a pin into a hasp or chain attached to a business or warehouse door. It is not unusual for a pin or sliding bolt to vibrate or "dance" out of the hasp or locking arrangement if an attempt is made to force the door with a burglar tool or pry bar. In earthquake areas, a door pin of this type may work out, apparently because of minor earth tremors.

If sliding bolts or pins are used, a dishonest employee can deliberately unlock a door just before leaving the premises, making it possible for a criminal associate to enter the building and steal after employees have left for the day. This would simply not be possible if the door were padlocked by the employee responsible for securing the building at the close of business.

Then, too, if there is a set procedure for padlocking each door on the perimeter of the building, it will be apparent in a routine inspection by management if the padlock has not been applied.

Experience indicates that roof burglars frequently drop down into an industrial or business building on a rope. Once inside, the criminal sometimes discovers that he is unable to climb back up the rope and that none of the doors in the building can be opened from inside. The intruder may still escape by forcibly breaking out of the building, providing he can find tools for this purpose. If the doors are securely padlocked from inside, the criminal may have no choice but to hope to escape when an employee opens the building.

Some professional burglars utilize a "hide in" technique, secreting themselves in rest rooms or storage areas of the building until all employees have left for the night. Here again, if all doors to the building are securely locked at closing time, the burglar may have difficulty in escaping or in getting merchandise out of the building. Quite obviously, doors that are openable from inside with thumb latch locks, sliding bolts, or pin arrangements are to the liking of "hide in" burglars.

Warehouse Doors

Any metal or wooden warehouse door that is unusually long should be padlocked on the inside at both ends, especially if there is exceptional play in the door. Experience also demonstrates that it may be desirable to padlock some doors of this kind in the center as well, making use of a floor-level hasp or a steel ring sunk in the cement floor of the warehouse.

A "hide in" burglar may also be able to raise an electrically operated warehouse or pedestrian door merely by pushing the button controls located adjacent to the door. These electrically activated doors can usually be secured with a lockable metal box control station that locks in place over the push buttons (see Figure 3).

A warehouse door that is raised or lowered by means of an endless chain should be secured by locking the door itself to the wall or floor of the building. Merely padlocking the endless chain to a fixed object may not prove adequate, as there may be so much play in the chain that the door can be raised several inches. Criminals may then be able to insert a pry bar, or a small individual may have enough space to crawl under the door.

Figure 4 shows how placing a heavy bar across the inside of a door may prevent a break-in. But the bar should be locked in place to prevent removal by someone already inside.

Door Hinge Pins

Exterior pedestrian doors frequently open outward, with the hinge knuckles or barrel on the exterior side of the door, as in Figure 5.

Utilizing a screwdriver and hammer or other hand tools, an intruder can drive the hinge pins out of some door hinges in a matter of seconds. If the door does not fit tightly in its frame, the door may be pulled out after the pins have been removed.

Physical Facilities 25

FIGURE 3 Lockable metal box control station for electrically operated doors.

Photo courtesy of Square D Corp., Milwaukee, Wisconsin.

FIGURE 4 A door bar provides better security when locked in place.

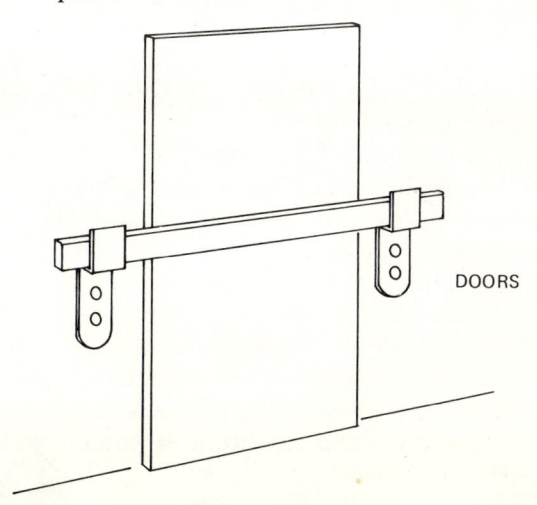

26 Physical Facilities

FIGURE 5 Exposed door hinge pins in need of protection.

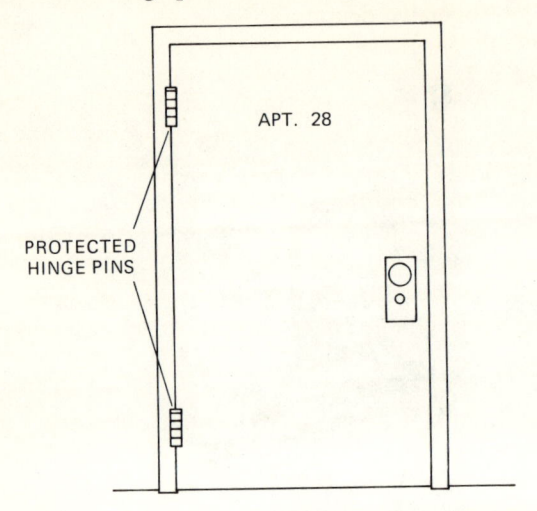

   In some installations, the door can be remounted so that hinge pins are inside the building, unavailable to would-be intruders. This kind of installation is sometimes impractical, and city fire codes often require all exterior doors to open outward.
   Exposed hinge pins can be protected in a number of ways. For example, pins can be spot-welded to the leaf of the hinge proper.
   Another protective technique involves inserting a headless machine screw in a predrilled hole through a leaf of the hinge and inward into the pin. Since the machine screw is inside the fold, it cannot be reached by a would-be intruder.
   Some mortise or "butt" type hinges are manufactured with a pin already extending outward from one of the leaves, made to fit snugly into a hole in the opposing hinge leaf when the door is closed. Carpenters sometimes complain, however, that this type of hinge is difficult to hang. This type of hinge is seen in Figure 6.
   Hollow-core wooden doors are seldom satisfactory for security because they are not usually strong enough for lock mountings. In short, they will not stand up to hard blows or prying tools. Experience also shows that an intruder may kick through a hollow-core door unless it is covered with a sheathing of sheet metal, completely covering the outer surface.
   Some pull-down type warehouse doors operate with small wheels mounted in a track at each side of the door mounting. These wheels are susceptible to breakage or damage, and the track may be bent out of shape by a blow from the warehouse forklift. Both the track and the wheels

should be kept in good repair, since the door can be removed if several wheels pop out of the track.

Some other pull-down type warehouse doors are constructed with inner panels of Beaverboard or pressed wood. Usually these panels are relatively thin and may be broken out with a length of pipe or a club. In some instances these panels have even been kicked through with a heavy shoe. Steel straps, about ⅜ to ½ inch thick and about 2 inches wide, can be bolted across the middle of each panel.

Round-head bolts, installed from the outer side of the door, will prevent easy removal of the steel straps or the retaining bolts themselves. Installing these straps makes the door more difficult to penetrate but also adds some additional weight that may make the door somewhat harder to raise.

FIGURE 6 Mortise type hinge with built-in pin to prevent door hinge removal.

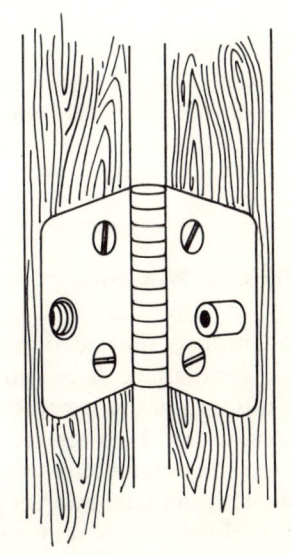

Windows

Criminals will usually attempt to enter a window in a rear alley or side street, to get away from a heavily traveled thoroughfare.

As a generalization, the higher the window the less likely that it will be used for entry. Windows with a ledge 18 feet or more above street level are seldom used by intruders. Criminals do, however, still enter high-level windows from adjoining buildings, or by climbing telephone poles or ladders.

Physical Facilities

Some lawbreakers will also attack windows along well-traveled streets, regardless of passing traffic. This is usually done in either of two ways:

1. By a "smash and grab" technique, the burglar running away from the scene.
2. By a stealthy attack with a glass cutter, the burglar using a small suction cup to prevent the glass from falling and making noise.

An alarm on the glass storefront window may alert the shop owner or the alarm company that the window has been broken. Before anyone can respond, however, the culprit may have snatched up jewelry or valuables and run away.

No window glass can be termed "unbreakable," but there are a number of products that are highly break-resistant. Installations of this kind greatly reduce the likelihood of loss to thieves using a "smash and grab" technique.

To frustrate the efforts of the stealthy criminal with a glass cutter, guard glass detectors are available. Made with an electromagnetic switch, these detectors may be tuned to precisely the frequency of vibration created by breaking glass. The devices are also sensitive enough to detect the use of glass cutters and still remain unaffected by normal shock or vibration.

Steel Mesh or Iron Bars on Windows

Many buildings have windows that need protection from intruders, yet must be openable for ventilation. It is usually advisable to cover first-floor windows at the minimum, using protective iron or steel bars or a good grade of heavy steel mesh. Bars are usually impressive from a security standpoint, but they often give a business a jail-like appearance.

In many respects, a steel mesh installation is preferable to steel bars. For example, a dishonest employee inside may be able to hand merchandise of reasonable size to a cohort outside, although the window is covered with bars. Mesh makes this impossible. Mesh likewise prevents the dishonest employee working alone from passing items outside, intending to return to pick them up later.

Mesh installations are also more effective than bars in preventing the introduction of fire bombs (Molotov cocktails) through building windows. In the so-called Watts riots in Los Angeles in 1965, Los Angeles police authorities reported no losses from fire bombs in buildings protected by mesh installations on both windows and doors. A number of buildings utilizing bars on windows were completely destroyed. Since 1965 there have been numerous other instances in which the protective value of steel mesh has been demonstrated.

Either bars or mesh should be permanently attached to masonry or steel window frames by welding or bolts. Mounted on wooden frames, either bars or mesh may be pried out or may loosen and fall away with age. Cases are on record of intruders who crawled through bars spaced only 5½ inches apart, although this is very exceptional.

Using Doors and Windows for Ventilation

Wire gates or covers should be considered for each door or window left open for ventilation. Experience shows that these gates or window covers may reduce theft losses considerably.

In many companies a maintenance employee can construct satisfactory wire gates or covers at reasonable cost. These may be made using standard pipe frames, with chain-link fence fabric welded or bolted to the pipes. Gates should be padlocked into place across door openings, with the fit close enough to prevent merchandise from being passed out.

Wire cages around windows will also allow ventilation while furnishing protection.

Painting the Inside of Warehouse Windows

From a security standpoint, warehouse windows should be painted on the inside. If a would-be thief cannot see the kind and quantity of merchandise that may be inside, there may be less inclination to break in. Painting windows reduces inside lighting level, but good inside lighting can be provided.

Window or Door Locks That Can Be Reached by Breaking Glass

Intruders frequently gain entry by breaking a glass pane in a door or window and reaching a thumb latch type lock inside. In situations of this kind, it may be advisable to substitute a solid-core door for a door with glass panels, or to brick up a window through which an arm may be inserted. Decorative steel mesh may also be placed on all glass openings that are likely to be broken out, and a lock may be installed that requires a key for opening from either side.

The Roof Hatch, Skylight, Air Conditioning Vent

Like a warehouse door, a roof hatch that is merely secured by a hasp and pin can be left unlocked by the criminal's accomplice who works inside the building.

Skylights and air conditioning vents may need protective steel bars or heavy steel mesh, if the opening is large enough for an intruder to crawl through. Designed to pop outward to localize a building fire, some so-called bubble type skylights should have protective bars or mesh installed underneath the skylight, rather than above. (See Figure 7.)

Generally, all openings on the roof should be included on an alarm system that utilizes contacts on the doors and windows of the building perimeter.

FIGURE 7 Skylights and roof vents should be protected with heavy mesh or bars, and connected to the alarm system.

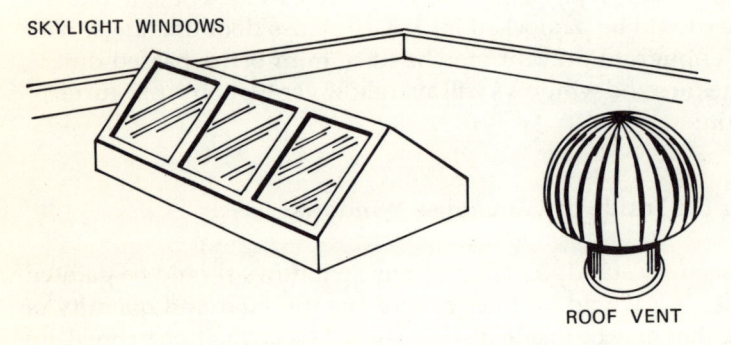

SKYLIGHT WINDOWS

ROOF VENT

Basement Access

In older structures it is frequently observed that cellar doors and basement-level windows are no longer used. These inherent security weaknesses can be corrected by permanently sealing the unneeded openings with reinforced brick work or masonry.

A meat packer in the Midwest operated continuously from one old building for many years. When space limitations became pressing, a new, well-planned structure was added to one end of the plant, but the old building was changed very little. A padlock was placed on a heavy steel coal scuttle door, located at the ground level in the outer wall of the old building. In some unexplained way, one of the keys to this padlock fell into the hands of a plant employee who had responsibility for transporting hams, bacons, and packaged meats inside the plant. Making use of the unauthorized key, this plant employee regularly passed substantial amounts of meat to a confederate in the alley outside the coal scuttle opening. These thefts apparently went on for over 20 years before discovery. The opportunity never would have existed if the coal scuttle opening had been closed with masonry, or if the heavy steel door on the scuttle had been permanently welded closed.

Ladders

Portable ladders should be brought inside a building, as they may be used by roof burglars. If the ladder must remain outside, it can usually be chained and locked to a tree or building fixture.

A ladder permanently affixed to the wall of a building may be made inaccessible by a steel cage around the steps, or by a long steel plate padlocked over the outer side of the ladder rungs.

Left for several days at a time, railroad boxcars are often spotted on a spur track immediately behind an industrial installation or a business warehouse. Ladders on these railroad cars can sometimes be climbed for roof access. Raising the height of the rear wall of the building may decrease this security risk. Or a chain-link fence installation at the edge of the building's rear wall may also prove effective.

If a forklift with sufficient capacity is left outside a building, one criminal may use this equipment to hoist another criminal to the roof of the building. This possibility, of course, is one argument for recommending the lift be maintained inside the building at night or when the business is closed.

Public Utility Openings

Underground openings for utilities are sometimes large enough to permit an intruder to crawl into a business or industrial building. Obviously, an opening of this kind may be another place where steel bars or heavy steel mesh should be installed.

It is also apparent that an unexpected disruption of electrical, telephone, or water services could completely halt operations of a business. A hasp, welded to a manhole cover and securely padlocked, may be enough to keep out strikers. In actual cases in the past, strikers have poured gasoline into underground telephone receptacles, fusing telephone connections by throwing a lighted match into the gasoline-soaked manhole.

SUMMARY

Physical controls, alone, are only part of a good protection program; they may merely deter or impede. But good obstacles to entry may convince an intruder that unauthorized entry may not be worth the effort or the chance of being caught. Generally, physical controls can be set up at three lines of defense: at the outer edge (perimeter) of the property, at the exterior of the buildings, and in specific areas inside buildings.

Landscaping, if properly designed, will not conceal activities in or around the building or approaches. Proper building placement will allow passing police patrols to observe what is happening on the property. In the plans for placement and usage of buildings, consideration should be given to perimeter controls, lighting, fencing, parking areas, and building entrance controls.

Chain-link fencing is perhaps the most commonly used type of perimeter protection, as it allows observation from either side. Good security fencing usually makes use of a "V" arrangement, made of metal arms carrying at least three strands of tautly strung barbed wire on each side of the "V." Fencing should be well anchored to the paving or ground, with gates as secure and as high as the fence itself.

Generally, physical security is no better than the protection afforded to doors, windows, skylights, and other openings in the building's outer shell. Windows and doors should be limited to the minimum number needed to operate the building. It is preferable for all doors but one to be padlocked from inside the building, with one control door in a patrolled, well-lighted place. Good locks are necessary on all doors and windows, with doors of sufficient strength to prevent lock hardware from being torn away. Door frames and installations must also be strong and rigid. It should also be kept in mind that some door hinge pins will need protection.

Generally preferable to steel bars, heavy steel mesh keeps inside thieves from passing out merchandise and keeps outsiders from throwing fire bombs through windows into the building's interior. Woven wire gates for doors and cages over windows are usually helpful in providing ventilation while giving reasonable security.

Skylights, roof hatches, air vents, and other building openings should also be protected with heavy mesh or bars in addition to being included on the perimeter alarm system designed to protect all building openings. If uncontrolled, ladders may be another threat to building security.

REVIEW AND DISCUSSION

1. Are physical controls alone sufficient to maintain security? Give reasons for your answer.
2. List three lines of defense on the property where physical protective devices or controls may be applied.
3. Give some general security rules for landscaping around a building.
4. Why is chain-link fencing the most commonly used kind of perimeter control? Did your answer consider the best protective feature of this kind of installation?
5. Identify two security defects that are frequently observed in chain-link fence installations.
6. What would you set as the minimum overall height of a good security fence?
7. Since gates are sometimes under the scrutiny of security guards, should gates have less height than the surrounding fence? Why?
8. Are most burglaries perpetrated by intruders who break a hole through a wall of the building?
9. Give a brief discussion of the techniques generally used by roof burglars.
10. From a theoretical security standpoint, what is the maximum number of doors that should be built into a business or residential structure? Why?
11. Why is it desirable for all doors in a business or warehouse to be locked from the inside except one?
12. What is meant by "pinning a door"? Why is this an undesirable security practice?
13. How do fire exit alarms work? May they be locked from the inside when employees leave the building for the day?
14. Why is it usually necessary to secure the hinge pins on an exterior pedestrian door? How may this be accomplished?
15. Comment briefly as to why an expensive locking device, alone, may not be sufficient for good door security. Why is a rigid door frame desirable?
16. Are hollow-core doors desirable for installation at exterior building locations? Give reasons for your answer.
17. Give a brief description of the common weaknesses of warehouse doors.

34 Physical Facilities

18. How high must a window installation be above street level to be reasonably safe from intrusion attempts? What should be done about windows below this level? Should you ever discount the possibility of intrusion, regardless of window height?
19. Describe two techniques used by outside criminals in stealing merchandise from store display windows.
20. Is any window glass really unbreakable? Give reasons for your answer.
21. How may security be maintained if doors or windows must be kept open for ventilation?
22. Why should warehouse windows be painted on the inside?
23. Should skylights, vents, and other roof or basement openings be protected as carefully as windows? Should all these openings be placed on the building's perimeter alarm system?
24. Are unsecured ladders a threat to security? Why?
25. Explain how building openings for the entry of public utilities may be sabotaged by a disgruntled employee or someone involved in a violent strike.

3

Locks and Keys; Safes and Vaults

Locks, safes, and vaults are an integral part of security. Their physical strength and resistance provide substantial protection, and their constant use provides a psychological deterrent to crime.

LOCKS AND KEYS

There should be a real lock on every exterior door, no matter whether it hinges, pivots, slides, rolls, folds, or lifts.

In selecting a lock, consideration should be given to physical strength, resistance to cutting, pounding, or picking, and duplication of keys. A good lock resists opening by any of these methods in varying degrees.

As part of the physical security system, the lock need not be any stronger than the door panel where it is installed, or any stronger than the hasp in which it is inserted. It also serves no purpose to use a lock of great strength if a duplicate key can be purchased in a nearby store.

Lock manufacturers frequently go to great lengths to stress features making their locks difficult to pick. Perhaps no single one of these devices is absolutely pick-proof, given enough time and attention. But there are a number of lock cylinders that are very difficult for the most experienced expert, and there are any number of other locks that are beyond the skill of amateurs. It is also worth noting that there are few real experts in picking lock cylinders, even among old-time locksmiths.

This means that the lock should be as resistant as possible, but there are other lock security requirements that are important. While statistics on lock picking are not available, authorities on burglary are in general agreement that criminals seldom enter by picking the lock. In a high percentage of cases, criminal intruders merely enter through openings that have been left unsecured. In most other instances, entries are made by forcible means, or by attacks against locking devices other than by picking.

If a glass door is locked, the burglar always has the option of breaking the glass. The noise and danger of breaking glass will usually scare the criminal, however, so other ways of gaining entry are generally attempted.

Cylinder Pulling

Standard lock cylinders are usually made of brass, with threads that strip easily when a heavy wrench is used to apply pressure to the exposed part of the cylinder. Many lock cylinders are also made with a shoulder that can be gripped or pried upon with hand tools. In some instances the burglar may carry a large set of portable tongs that actually pull or rip out the cylinder through the front of the door.

Some locks are manufactured with the outer cylinder ring of beveled, case-hardened steel. A hardened steel cylinder guard or ring with a steel retainer plate located on the inner side of the door will make it extremely difficult to pull the cylinder. This type of installation is seen in Figure 8.

Springing the Door

If a door lock does not have a deadlock bolt that projects at least ¾ inch into the strike hole (the bolt receptacle), it can frequently be sprung without leaving visible signs of forcible entry. Aluminum doors, of the type used in modern store fronts or professional buildings, are especially vulnerable in this regard. Quite flexible, door frames of this type bend enough to free the bolt when pressure is applied with heavy screwdrivers or pry bars, as seen in Figure 9.

This type of locking deficiency can usually be corrected by using a lock with a long, 1⅜-inch pivoted bolt of laminated steel.

Locks and Keys; Safes and Vaults 37

FIGURE 8 Cylinder guard of case hardened steel and retainer plate prevents cylinder pulling attack.

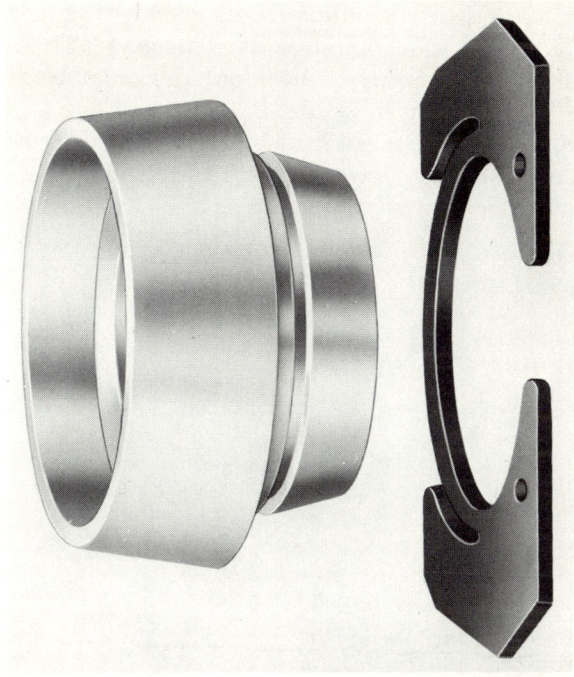

Photo courtesy of Adams Rite Manufacturing Co., City of Industry, California.

FIGURE 9 Springing the door frame to allow the bolt to swing free.

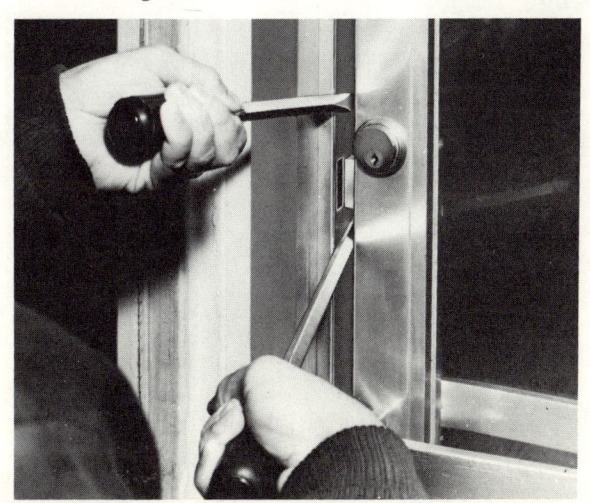

Photo courtesy of Adams Rite Manufacturing Co., City of Industry, California.

Jamb Peeling

One method for getting around a deadlock bolt of sufficient length is by using the burglary technique called "jamb peeling." This involves use of a pry bar to tear away enough of the door frame or jamb so that the bolt swings free. This is shown in Figure 10a.

Jamb peeling techniques can usually be prevented by using a heavy steel door frame, or by installing an armored strike plate of steel, covered with a decorative face, as seen in Figure 10b.

FIGURE 10 (a) Jamb peeling technique and (b) use of an armored strike plate.

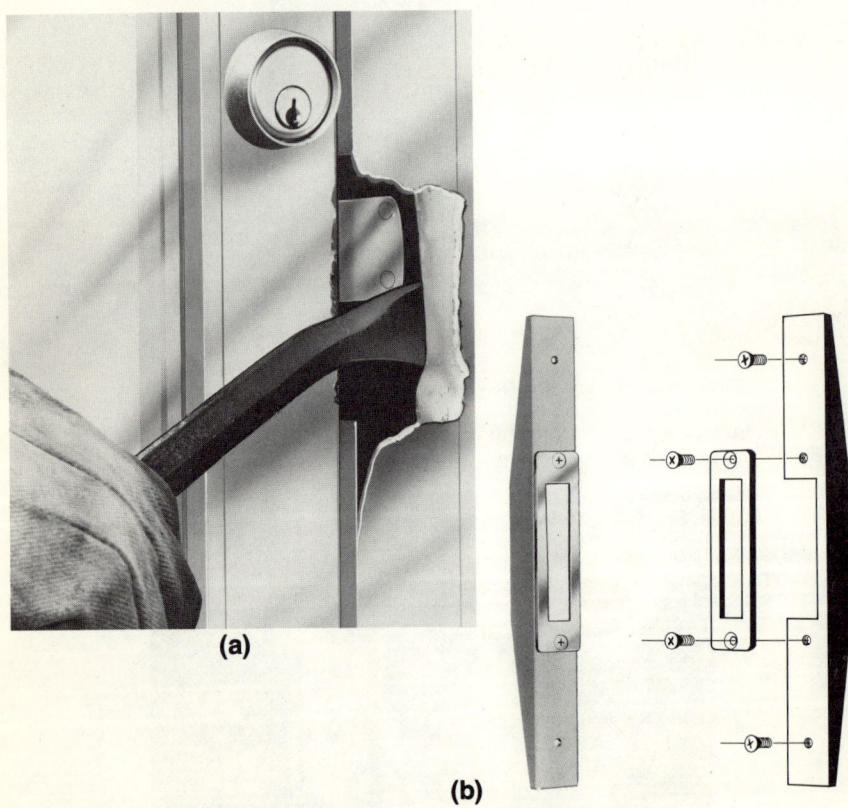

(a)

(b)

Photo courtesy of Adams Rite Manufacturing Co., City of Industry, California.

Bar and Hammer Attacks, and Sawing the Bolt

Other burglarly techniques sometimes involve attempts to drive out the door bolt with a hammer and bar, or to saw through the bolt with a hacksaw blade, as seen in Figures 11 and 12.

Locks and Keys; Safes and Vaults 39

FIGURE 11 The hammer and bar technique to force out the lock bolt.

Photo courtesy of Adams Rite Manufacturing Co., City of Industry, California.

FIGURE 12 Sawing the lock bolt with a hacksaw.

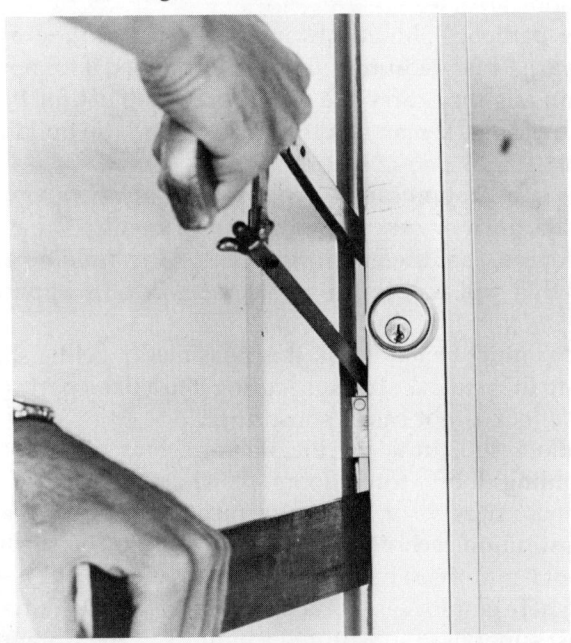

Photo courtesy of Adams Rite Manufacturing Co., City of Industry, California.

Modern lock manufacturers have made use of a pivot lock bolt to resist the hammer and bar type of attack. Further, an alumina-ceramic core in a laminated steel bolt has been devised which makes sawing the door bolt an extremely difficult and time-consuming method of attack; the device stops even a tungsten carbide rod saw.

Padlock Substitution

Padlock substitution is a criminal technique used to gain access to a door where the padlock is left unsnapped during the daytime. The thief begins by obtaining a padlock that is similar in appearance to the lock hanging at the door. While no one is observing, the criminal will pocket the unsnapped building padlock, leaving his own padlock in its place. When the door is secured for the night, the building employee will automatically snap the substituted padlock in place.

Later, when the thief desires to open the locked door, he will take out his key and open the padlock which he substituted. After taking whatever is desired, the thief removes his own padlock and snaps the building lock back into place. When the building keyholder returns on the following morning and opens the building padlock, he has no way of realizing that anything is wrong. When the theft is discovered, all the known facts will indicate that the theft was internal and thus only building employees could be responsible.

Building padlocks should always be locked around a fixed object (snapped shut). This precaution should be followed if for no other reason than that vandals may carry the padlock away "just for the hell of it." With a lock missing, it may be difficult to secure the building at closing time.

Padlock substitution can be prevented by always snapping the lock shut around a hasp or other physical attachment to the building. This means, of course, that the employee locking the building at night must carry a key that will unlock the padlock prior to its application on the door.

If the lock must be left open, it is possible to weld a short length of small chain to the padlock shank, securing it to a fixed part of the building door. Such a lock is not easily substituted.

The padlock substitution principle applies to all padlocks, whether used on building doors, truck doors, or whatever.

A dishonest supervisor in a major meat packing company used the padlock substitution technique to steal over $100,000 in choice cuts of meat in about 6 months. The meat was taken from a frozen storage room in a facility where employees worked 5 days a week. On Friday afternoon, the dishonest supervisor substituted a padlock at the storage room door.

Every Saturday he opened the lock and used a company fork-lift to remove large quantities of meat. Eventually, a suspicious manager concealed himself in an inconspicuous spot and observed what was happening when the plant was closed for the weekend.

The fact that a padlock is used on the inside of a warehouse door does not mean a padlock substitution scheme cannot be used. Generally, of course, a thief will substitute a padlock on a door that can be reached from outside the building. In one recent case, a burglar substituted a lock that was used on an inside hasp in a spice warehouse. The burglar then hid out in the building, emerging from concealment after all employees had left for the day. Opening the door that had been secured by the substituted lock, the criminal then removed exotic spices valued at about $17,000.

Protecting Hasp Installations

It is sometimes observed that padlocks are used on hinged hasps. The padlock itself may be of good quality, but security may be compromised by improper installation of the hasp. The outside leaf of the hasp should fold back over the inner leaf, covering the heads of the wood screws or bolts that secure the hasp to the building or door. If improperly installed, the screw heads can be quickly removed with a screwdriver. A proper installation, with no exposed screw heads or bolts, is seen in Figure 13.

FIGURE 13   A hinging hasp with no exposed screwheads.

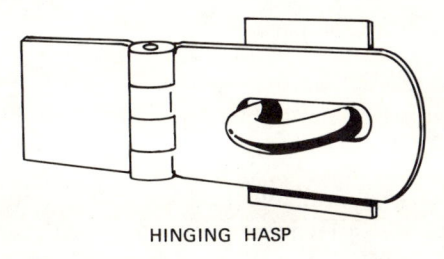

HINGING HASP

Removable Core Locks

Removable core locks offer flexibility that is not available in other types. In the event a key is lost or stolen, the core can be changed in a matter of seconds, without the services of a locksmith. See Figure 14. This replacement, in effect, has made a new lock of the old device.

Costing somewhat more at the time of purchase, removable core locks usually permit security to be maintained with less replacement cost. Applicable for use in building doors, this type of lock can be used for most business and residential installations.

42   Locks and Keys; Safes and Vaults

FIGURE 14   A removable core lock.

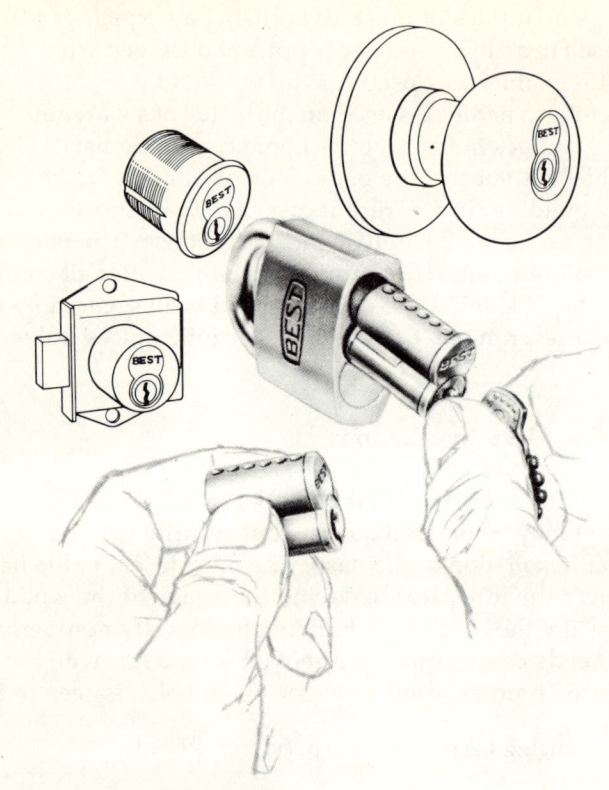

Photo courtesy Best Lock Corp., Indianapolis, Indiana.

Specialized Locks

Specialized locks are available for a number of unusual applications. Some padlocks, for example, are manufactured with case-hardened steel shackles, to resist sawing or cutting with a bolt cutter. Others are designed to withstand considerable pounding with a heavy hammer. In still another type, the key cannot be removed until after the lock has first been snapped into a locked position.

Locks may be keyed alike or keyed differently. Some manufacturers will provide special keyways, which are assigned to an individual customer and are not furnished to locksmiths or other key duplicators.

Fifth-wheel or kingpin locks are out-sized devices designed to protect a detached freight trailer. Essentially, this kind of lock is a metal boot, made to fit over the fifth wheel or kingpin on a trailer, so that a truck cannot be hooked up to pull the trailer away.

Locks and Keys; Safes and Vaults   43

Combination locks have a number of applications, with a wide range of quality. The combination lock on a well-made safe or bank vault may offer excellent security, whereas an inexpensive combination padlock such as that used on a student's school locker may not.

Pushbutton locks and card-activated locks are also widely used in some applications, such as on the entry door to a computer room installation. See Figures 15 and 16. Here again, quality varies, but some of these locks provide good security. They are seldom adequate for exterior doors.

With a cardkey lock, the same card can be coded to allow entry to more than one door. The combination can also be changed on site, without the aid of a locksmith.

In more sophisticated applications, the cardkey system can be teamed up with a computer. When it is necessary to invalidate a card, the system can be programmed to rule out entry in a matter of minutes. The computer can also maintain a record as to when an individual card was used to open a specific door. This information can be very significant in determining who had access to a restricted area where a crime or security breach occurred.

FIGURE 15   A pushbutton or digital combination lock.

Photo courtesy of Preso-matic Lock Co., Fort Pierce, Florida.

The Recording Lock or Time Lock System

Unexplained losses in a retail or wholesale business can sometimes be traced to keyholders who steal from a business after it has closed for the night or weekend.

44 Locks and Keys; Safes and Vaults

FIGURE **16** A cardkey lock.

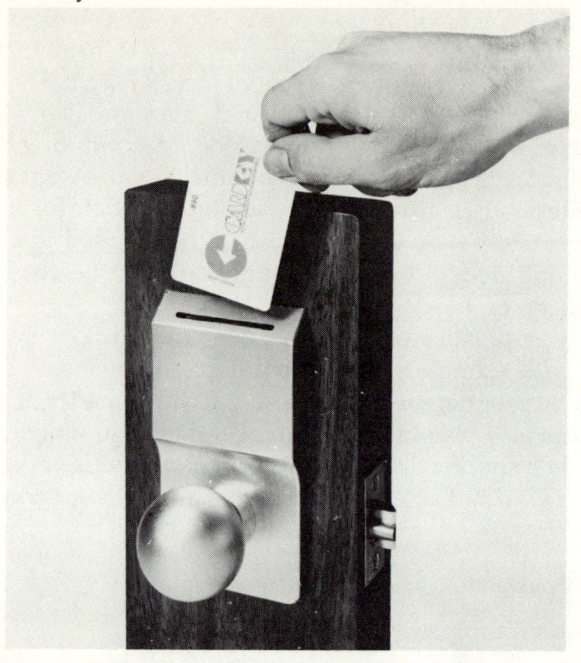

Photo courtesy of Cardkey SYSTEMS, Chatsworth, California.

Special kinds of locks are available to business owners or managers who want to be aware of opening and closing times, as well as reentries after normal working hours by key-carrying personnel. A locking system of this kind is known as a recording lock or time lock. Essentially, these devices make use of a timing mechanism and a recording system in a small locked box, attached to the building's control door. The act of locking or unlocking any exterior door sets the timing and recording units in motion, printing out the date and time, which door was opened or closed, and the key used by letter designation. The record from the locked box is regularly removed and mailed to management, preferably at a home address where it cannot be intercepted by employees.

This type of control device may include as many as five doors and five keys. In addition to building exterior doors, the date and time of opening by specific key can be recorded for any of the following:

Raising and lowering of overhead warehouse doors.
Opening and closing of all delivery doors.
Opening and closing of emergency exit doors, whether by key or by depressing the panic bar and setting off an alarm.

Opening and closing of safes and vaults.
Opening and closing of stock or storage rooms.
Opening and closing of computer room doors.

A typical printed record from a recording lock system is shown in Figure 17.

Much of the value of a recording lock or time lock system is psychological, since keyholders are unlikely to make unauthorized openings that must be accounted for.

Recording lock and time lock devices can be used with a sequence locking system. This arrangement makes certain that an employee will not forget to lock a building door, since the front door lock cannot be set unless all other doors on the system have been locked in sequence.

FIGURE 17 A recording lock system.

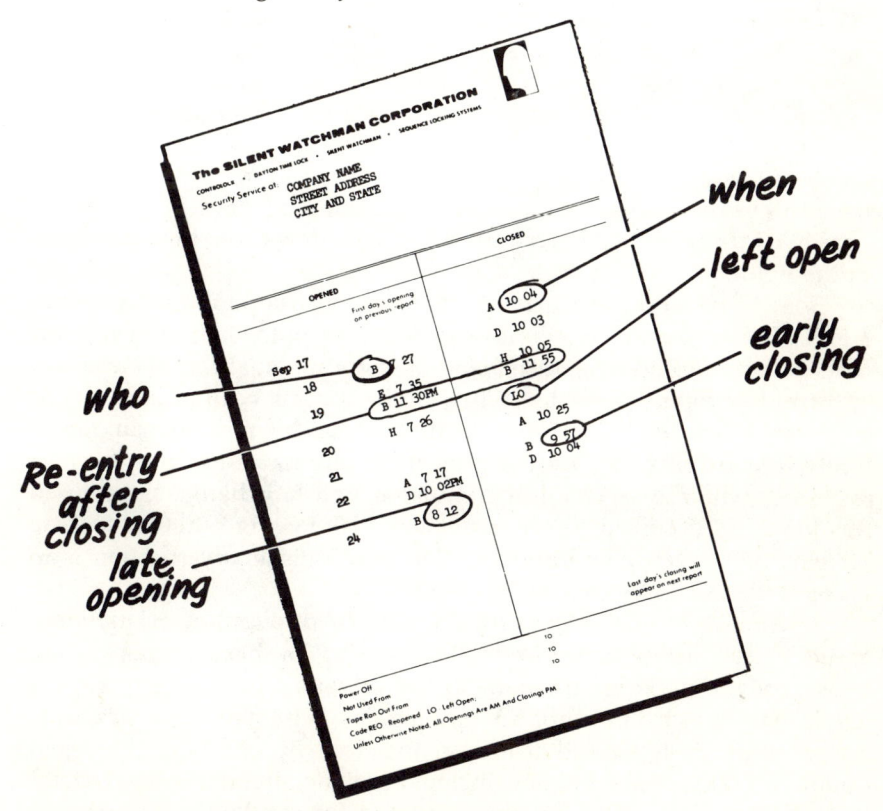

Photo courtesy of The Silent Watchman Corporation, Columbus, Ohio.

Key Control

Whatever locking system is used, it is no more secure than the key control procedures that are used. If a key is missing or unaccounted for, it must be assumed that it may have come into possession of an unauthorized person.

Perhaps the first principle of key control is to minimize the number of outstanding keys. If an analysis is made, it will frequently be found that keys to a business or organization have been issued solely for prestige, to persons who have no real need. When a New York firm recalled outstanding keys, it was found that 28 had been issued for the front door to a one-story building less than half a block in area.

The company employed a number of salesmen who claimed that they needed entry to the building on weekends. The stockroom supervisor, the head of office services, the company accountant, the warehouse superintendent, three computer room supervisors, and the head of personnel all insisted that they also needed keys. Five members of management held keys because of job necessity. One of these five managers was always at the building, including on weekends. Perhaps these five were the only door keys that should have been issued.

There was no certainty that additional keys had not been passed out over the years. The great number of outstanding keys increased the possibility that keys had been duplicated, with or without the knowledge of the holders.

Sometimes a business manager or warehouse superintendent hands a ring of stockroom or warehouse keys to an employee, instructing the latter to open up merchandise areas. It is not unusual for an employee entrusted with the keys to return them several hours later. If this employee keeps the keys during the lunch hour, he may have an opportunity to copy each key on the ring, or to duplicate sensitive keys for associates who are involved in criminal activities. Where a situation of this kind has occurred, there is no way of knowing whether locking systems have been compromised. The only satisfactory solution is to make certain that keys are never loaned.

A somewhat similar opportunity for key duplication exists when an individual drives his car to a public parking lot that requires keys to be left for the parking attendant. If the driver's whole ring of keys is left in the switch, a dishonest parking lot employee may duplicate house or business keys that are on the ring. By checking the registration in the car, the criminally inclined parking attendant can discover the address where the keys may be used for residential or business access.

Key Audits as a Control

A regular audit of business keys may be a vital part of control. An individual holding a key may sometimes be afraid to report loss, or may actually be unaware that a sensitive key has been taken. The employee who has been issued a key should be asked from time to time to produce it, and the key should then be tested in the lock to make certain there has been no error. At the time of an audit it will sometimes be found that a sensitive key has been mislaid, loaned, or lost. The auditor may also observe that the key is maintained in an unlocked desk or other location where unauthorized persons have access to it.

Precautions Against Key Duplication

Management is never quite certain whether an employee has allowed another individual to have an unauthorized key made. It may therefore be advisable to have a company rule stating that unauthorized duplication of a company key makes the offender subject to dismissal.

Some companies have gone to the trouble of using only specially designed key blanks with the company name on the bow. Others feel that the company name or identification should never appear on a key, for fear that a person finding such a key would know where the key could be used in an illegal way.

Some firms place the notation "Do not duplicate" on all key blanks. Reputable locksmiths will refuse to duplicate a key that is marked in this way, but there is no certainty that this restriction will be followed.

If special key blanks are used, an audit by a company official will make it obvious if a duplication has been made on an ordinary blank such as a local keymaker would use.

In some situations, security consultants recommend that serial numbers on the base of padlocks and on keys be scratched off by a few strokes with a common hand file. This is because duplicate keys can usually be ordered by furnishing this number and the type of lock to a locksmith. Unless issuance records are properly maintained, however, management may have considerable difficulty in identifying both keys and locks from which serial numbers have been removed.

Security for Duplicate Keys

Failure to control duplicate keys is a common security weakness. In a small company, duplicate keys may be retained in a locked drawer in a safe. It is not unusual, however, to find duplicate keys thrown into a cigar box, with little concern for identification or security.

48 Locks and Keys; Safes and Vaults

Another problem here is that the custodian of duplicate keys may have no definite instructions as to which company employees are entitled to use duplicate keys. For example, a large meat packing firm in the South carefully identified each duplicate key, maintaining these keys in a locked steel cabinet in the guardhouse at the front entrance to the property. Only a few officials had been issued sensitive keys. The guard, however, allowed anyone with an employee's badge to pick up duplicate keys, since the guard had never been told who had authority to obtain the backup keys.

Maintaining Key Records

A positive key system is needed in almost any company or institution for the issuance of keys. To be protected, records should be maintained both for keys issued to individual employees and for keys outstanding to each lock or door to be controlled. When security requirements for a section or department are changed, then keys should be reissued according to need.

When an employee is transferred or terminated, the personnel file should be reviewed to determine the continuing need for keys. Many companies require an exit interview with all departing employees, to make certain that employee keys have been turned in.

It is suggested that an individual receipt be issued when a key is given to an employee. A typical receipt form is shown in Figure 18.

FIGURE 18   A typical receipt form.

```
                         KEY RECEIPT

   I have received this date a key to_____facility.
   This key is property of _____company
   and must be returned on request, transfer to another job,
   or when leaving the company.  Any abuse of this key,
   unauthorized loan or duplication may be grounds for dis-
   charge, and I understand my responsibility in this regard.


                                  Signed _____


   DATE OF ISSUE _____Key #_____


   EMPLOYEE NAME _____ Payroll #_____.


   DEPARTMENT WHERE ASSIGNED _____
```

Locks and Keys; Safes and Vaults   49

A detailed key control work sheet is needed in almost every business. If a firm or company building is of unusual size, it may be necessary to use a number of key control record sheets, by building, floor, or section. In using a work sheet of this kind, a specific entry should be made for each key issued.

A key control sheet of this kind is shown in Figure 19. Individual notations should be made in the vertical columns for each lock or door

FIGURE 19   A key control work sheet.

	Front Door Main Bldg.	Freight Dock #1	Freight Dock #2	North Warehouse	Main Stockroom	Production Loft	Emergency Fire Door	South Entry	Parking Lot #1	Parking Lot #2	Supply Room	
Edward King - Engineer	X											
Will Henry - R&D			X	X			X			X		
M.K. Smith - Production					X		X					
Jim Henley - Sales		X			X				X			
Ned Ray - Chemist	X											
F. Lindley - Janitorial	X						X	X				

that is to be controlled. Horizontal columns should be used for each employee who receives a key. Then an "X" may be placed in the square where the specific lock and the employee's name coincide.

Reading across the work sheet (horizontally) will show which keys have been issued to a specific employee. Obviously, these are the keys from that section which must be returned in the event of an employee transfer or discharge. Reading downward in the vertical columns will show which employees hold keys for a specific door or lock.

SECURITY FEATURES
OF BUSINESS SAFES AND VAULTS

There are three basic types of safes, designed to meet the differing needs of business:

1. The money safe, or strongbox.
2. The business record safe.
3. The computer record safe.

Numerous cases are on file to show that reliance on the wrong kind of safe may result in serious loss. The importance of the differences among safe types should be understood, not only by security people but by everyone in business management.

Names of the three kinds of safes are descriptive. A money safe offers good protection against the burglar or robber intending to carry away currency or valuables. In the event of an extensive fire, however, the contents of a money safe may be roasted.

A business record safe protects vital company books and records against fire but may be easily ripped open by a burglar with proper tools. Keeping money in this kind of safe is inadvisable.

Neither a money safe nor a business record safe will give the necessary protection for computer tapes or disks. It is to be emphasized here that protection of bookkeeping records or computer records may be far more critical to the success of a business than safeguarding receipts, although all deserve good security measures.

Still another type of security installation is the walk-in record vault. Equipped with a steel door and a combination dial lock, the walk-in vault is frequently used for protection of either business bookkeeping records or company money, or both.

The money safe is made of heavy steel plates, tempered to withstand drilling, cutting, or pounding with sledges or heavy tools. There may be a "relocking device" inside the mechanism of a safe of this kind, causing the bolts to automatically relock the door if the spindle is punched out by a so-called punchman or knob knocker.

Essentially, a business record safe is of double-walled construction, with both inner and outer walls of relatively thin steel. The space between these walls is filled with anhydrous powdered chemicals, materials that turn into steam when fire reaches the outer walls. This steam vapor serves as a protective barrier, sealing off the cracks around the door and insulating the walls. Edges of business records inside the safe may be dampened by water vapor, but records will usually remain usable even after the safe has had considerable exposure to fire. An Underwriters' Label (UL) inside the door of a fire safe usually specifies the protection that can be expected in case of fire—a 2-hour rating, 4-hour rating, etc.

Once decomposed by exposure to heat, the chemicals between the walls of a record safe have lost their effectiveness. Therefore, a reconditioned safe of this type cannot be relied on for protection against fire.

Some business safes are built to incorporate the protective features of both the money safe and the business record safe, so that both money and records can be stored. These combination safes are usually more expensive and larger in size than individual types.

Computer safes fill a third business need. Computer tapes and disks will usually be ruined by even short exposure to water vapor at 130° to 150° Fahrenheit. Since water vapor is harmful to computer records, a computer safe is constructed of special insulating materials and also has strong steel walls to resist forced entry. This kind of safe is usually a cabinet type, large enough to hold a practical quantity of tapes or disks.

There are many techniques that burglars use to get into safes. The more common methods are described in the chapter on burglary (Chapter 10).

A number of protective features can be added to a money safe. A time lock attachment precludes the possibility of opening the box until the set time has expired.

Some money safes are made to open with either of two combinations. One of these three-number combinations will open the safe and at the same time send a silent alarm to the police or to a central station alarm company. Another device of this type is activated by dialing a prefix before the three-part combination. Other money safes are opened by keys, rather than a combination lock. Some require the insertion of two keys at the same time, with the keys issued to two different employees for better control. Still other safes are opened by a combination dial, with a key-controlled money compartment located inside. Such safes may have several separate locked compartments inside the chest, so one employee will not have access to money or records entrusted to others.

If both money and checks are kept in a safe or vault, it may be advisable to follow the practice of separating the checks from the currency. Unless this is done, a criminal may carry away the checks with the money, later discarding the checks. When both are taken from the premises, the business is seldom able to reconstruct the checks or to ask customers for replacements.

Selecting a Location for the Money Safe

Opinions differ on where a money safe should be located. Feeling it is better to conceal the fact that a safe exists, some small firms place the money box in an inner office, where it cannot be observed by any outsider. But most people know that almost every business has a safe. If a burglar once locates the concealed box, he is in a good position to devote his efforts to opening it with little risk of observation. Usually, hiding the safe serves only to give the burglar better working conditions.

The consensus of opinion in security is that the money safe should be located in a well-lighted room that is immediately in front of a window within full view of everyone on the sidewalk or street outside. In a private home, however, it may be preferable to conceal the existence of a safe behind a secret panel or inside a closet.

There is at least one drawback to making the business safe visible from the street. An experienced, professional burglar will then know whether it is within his technical ability to open this class of safe. It is unusual for a criminal to attack a safe located within view, unless the strongbox can be quickly rolled out of sight.

Police officers sometimes point out that a business safe has been placed in front of a large window, but that office employees have drawn the curtains at the close of business. Employees should therefore understand the importance of keeping the safe visible.

It is recommended that good lighting be placed above the safe. See Figure 20.

If a light is unaccountably out over a business safe, there is always a possibility that a burglary is in progress. Training for police officers and

FIGURE 20   Good lighting is effective in protecting the safe.

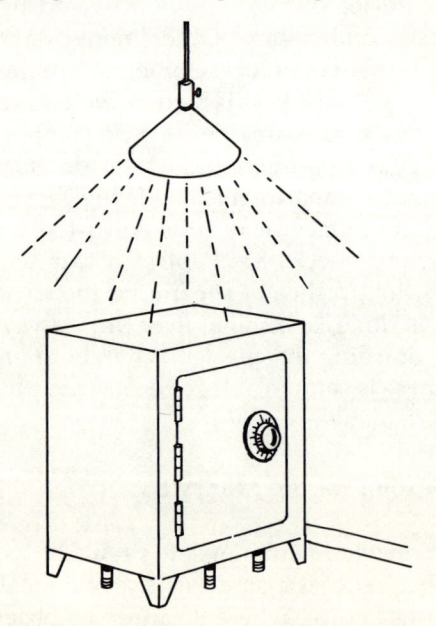

private security guards should point this out as a situation indicating an immediate investigation, with a cautious approach.

If an outside light is used to illuminate a safe inside an office, then precautions should be taken to protect the light installation. Break-resistant fixtures or a covering of heavy steel mesh may be utilized.

Making Certain the Money Safe Is Not Easy to Carry Away

A good safe may be very difficult to open, even for an experienced professional burglar. However, if the strongbox can be transported to an isolated farmhouse or a hidden ravine, the safe can eventually be opened by inexperienced criminals who batter the box repeatedly with a sledge hammer. It is therefore important to remove casters or wheels from the safe. See Figure 21. An even more effective precaution is to bolt the safe to a concrete slab, or to install it into a poured cement floor.

FIGURE 21   Removing the wheels makes it difficult to haul off the safe.

If a business safe is located in the floor, employees may be able to conceal the location of the safe by throwing a rug over it at the time the firm's office is closed.

A favorite technique of burglars has been to roll the safe from the business office to a company freight dock or loading area and then push the safe into the bed of a pickup or flatbed truck. If the safe is extremely heavy, it may break through the floor of the truck. If the burglars can get the safe away from the premises, however, they are under no pressure to open the safe within any given time. Some criminals refer to this type of burglary as kidnapping the safe.

This idea of making the safe immobile should also be applied to business cash registers, especially where there is a possibility that a criminal may unexpectedly grab a small register and run away. At times the store clerks may be too startled to respond to this kind of situation.

Controlling Tools and Equipment That May Be Used Against the Safe

If a company maintenance man uses a welding torch, heavy drills, pry bars, and heavy cutting tools, it is recommended they be secured when not in use. Knowing that this kind of equipment is on hand, a burglar may find his task considerably easier.

Special Safes for Delivery Vehicles

A company route driver who collects cash may need protection against armed robbery on the route, as well as a drop box for collections after return to the company premises. If the route driver is delayed, all supervisors may have gone home before the arrival of the delivery truck.

Small, specialized safes are manufactured for trucks, small stores, theaters, and other money collectors. Some of these may be no larger than 4 by 6 by 10 inches; they are usually bolted or welded to the truck frame, or a steel beam in a building installation. If a safe of this kind is installed in a delivery vehicle, a sign above the safe should state clearly that the driver has no key. Some safes of this kind can be opened only by the insertion of two keys, with the driver carrying one and the company supervisor holding the other. An example of this kind of safe is pictured in Figure 22.

FIGURE 22 A money drop steel safe that may be used in a delivery truck.

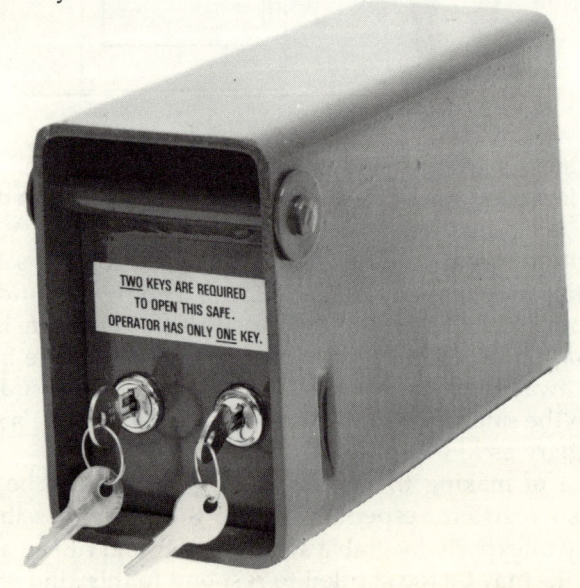

FIGURE 23   A typical "day latch" in a business vault.

Walk-in Vaults

Businessmen often assume that a walk-in vault offers more protection than is actually furnished. The dial combination steel door itself may give good protection against burglars as well as against fire. Most doors of this type carry a designated Underwriters' Label (UL) fire rating. But often the sides, floor, or ceiling of the vault may be considerably easier to penetrate.

In a well-constructed vault, the entire shell may be of concrete poured into a steel-reinforced framework of good thickness. It is not unusual to observe poor wall or ceiling construction, however. In some instances, the ceiling may be of sheetrock or of accoustical tile. Regardless of the protective qualities of the door, the vault may be penetrated through the ceiling crawl space from an adjoining office. Obviously, a vault of this kind does not adequately protect against either fire or burglary.

Many vaults are equipped with an inner door or gate steel grillwork or an aluminum grid; this is called a "day latch" or "day gate." Lockable with a key, these inner gates are designed to provide some security with reasonable convenience. From a practical standpoint, it is usually easier to use the day latch than to work the combination for the outer door. A typical day latch is pictured in Figure 23. Experience shows that businesses frequently become lax in locking the inner door, unless management insists that this be done.

If a company vault is relied on to protect computer tapes, disks, or other essential business records, it may be advisable to keep the outer vault door closed during the daytime. Experience from actual cases shows that employees may rush from the building in the event of fire, without taking time to go close the fire-rated outer door on the vault.

Using the Protective Features of the Safe

Many cases in police department files indicate it may not be necessary for a burglar to have technical ability in order to get into a business safe. Forgetful employees often write the safe combination on a desk blotter, on a card inside the top drawer of a desk, or in some other obvious place. A security consultant recently found a combination written in red crayon on the back of the safe.

One expert locksmith always began a lecture on safe security with this advice, "Always try the handle, the safe may not be locked!" A definite closing procedure should therefore be routine in any business using safes, vaults, or security-type file cabinets.

It is also suggested that the combination to the safe or vault be changed immediately after an employee severs business connections with the firm. A change of combination is usually justified, even if an employee is transferred to another company location. In addition, some security directors request the combination be changed on an annual or biannual basis, on the possibility that the combination could have been compromised without knowledge of the employees in possession of it.

In some businesses, the safe or vault is opened immediately after arrival of the first employee who knows the combination. This may be necessary if the safe or vault contains records or bookkeeping materials that are needed by other employees. If there is no immediate need for the items inside the safe, consideration should be given to delaying the opening. The probability of an armed robbery is increased by having the money chest or vault open either before or after the hours when access is actually needed.

Insulated and Lockable File Cabinets for Security

Insulated and/or lockable file cabinets may give reasonable protection against either fire or theft at prices considerably below the cost of a safe or vault. If records are classified according to replacement value, security priorities may dictate that some records be protected in a locked or insulated cabinet, rather than in a safe or vault. A practical locking device of this kind is seen in Figure 24.

Locks and Keys; Safes and Vaults   57

FIGURE 24   An insulated and/or lockable file cabinet that provides reasonable protection at less cost than a safe or vault.

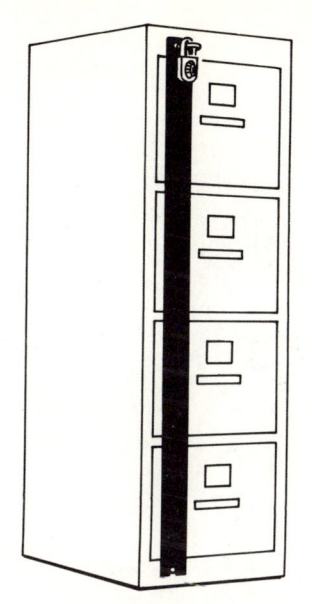

Lockable Cash Drawers

Lockable cash drawers are frequently used to good advantage in retail stores and shops, or in locations where the clerk cannot remain at the cash box. If the drawer is not consistently locked, it is usually advisable to have a bell arrangement that signals the opening of the drawer.

Some firms have found that placing a spring behind the drawer is a useful security measure. The spring forces the drawer out when it is not locked. The open position of the drawer is then readily observable, indicating immediately that the clerk has failed to lock the box.

It is generally advisable to put cash drawer funds into a safe or vault at the close of business. Even a steel drawer of the kind seen in Figure 25 will not withstand much prying from a bar or even a large screwdriver.

SUMMARY

Locks provide both psychological and actual protection. Ideally, there should be a lock on every exterior door, regardless of how it closes. But it is pointless to use a good lock if it is mounted in a weak door, or if the door

58 Locks and Keys; Safes and Vaults

FIGURE 25   A lockable metal cash drawer, with spring mounted behind drawer.

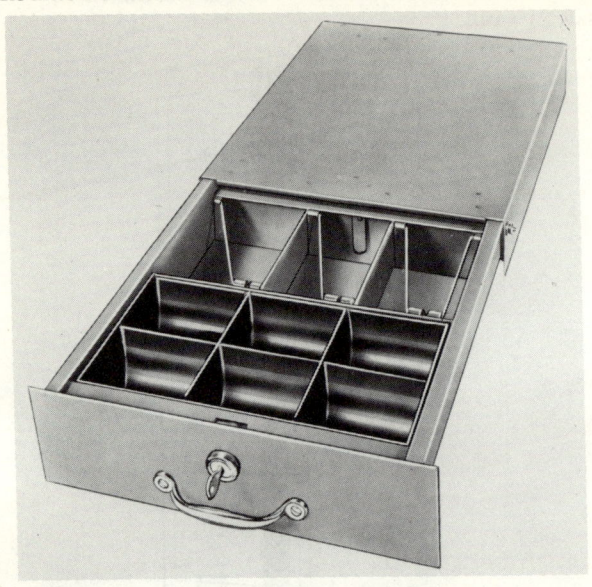

Photo by Indiana Cash Drawer Company, Shelbyville, Indiana.

frame can easily be broken open. As a practical matter, a lock on a glass door may not be adequate security.

Locks are often compromised or broken out by cylinder pulling, by springing (separating the door from the frame), by jamb peeling, by bar and hammer attacks, or by sawing the bolt. Burglars usually enter after a physical attack on the lock, rather than by picking the mechanism.

Usually, padlocks should be left snapped (locked) to avoid padlock substitution by a thief. Hasps should be installed so that screws cannot be reached for easy removal.

Removable core padlocks usually provide good security at less expense than replacing the padlocks entirely. Push button locks, combination locks, and cardkey locks all have useful applications in security.

Time lock systems provide an exact record as to when an individual key was used to make an after-hours entry.

Restricting the number of available keys and a system for key control are both vital requirements in a good security system. Regular audits of outstanding and duplicate keys are recommended.

Money safes are manufactured of heavy steel to resist attacks by burglars. A money safe may offer little protection against fire, however. A fire safe protects essential business records from fire, but may not be adequate to keep out an experienced burglar. A safe that combines the features of both the money safe and the fire safe is preferable. Computer records need still a third type of safe—a specially constructed computer tape or disk safe—because of unusual characteristics of these tapes or disks.

Safes should be located under a good light, so police and passers-by can readily observe persons attacking the safe. It is also desirable for the wheels to be removed from the safe, with the strongbox bolted or secured to the building floor or steel struts.

Walk-in vaults and insulated and lockable file cabinets and cash drawers may also be used for additional security protection.

REVIEW AND DISCUSSION

1. Explain why security representatives usually feel it is necessary to have a lock on every door, whether it slides shut, comes down from overhead, or closes in some other manner.
2. Explain the statement that it is pointless to use a good lock unless it is installed in a solid door. Why should hollow-core doors not be used for security protection? Outline why the door mounting must also be substantial, with the door frame of good construction.
3. What is meant by the term "cylinder pulling"?
4. Describe how a burglar may enter by "jamb peeling." What is a "bar and hammer" attack on a door lock?
5. Do thieves usually get through a locked door by picking the lock? Explain your answer.
6. What is the technique that thieves call "padlock substitution"? Outline how this works.
7. Explain the use of removable core padlocks.
8. How does a time lock system identify when an individual key was used to open a specific door? How may the system be used in security applications for business?
9. Sketch out or describe a good key control system to account for the issuance of outstanding keys.
10. What are the basic differences in money safes, fire safes, and computer safes? May a combination fire and money safe be used?
11. Why should there be a good light over a company safe?
12. What can be done to prevent a safe from being rolled away by burglars for loading in a pickup truck?
13. Explain the security features of walk-in vaults, insulated and lockable file cabinets, and lockable cash drawers.

4

Problems of Maintenance and Janitorial Access

The purpose of this chapter is to analyze security problems involved in maintenance and janitorial access to buildings and property. This relates to ways in which company equipment and tools may be protected, and the theft risk reduced in disposing of trash. The last part of the chapter is concerned with security through fire protection, examining causes of fire and methods for suppressing it. Automatic water sprinkler systems, hand fire extinguishers, hose systems, fire escapes, fire doors, and fire detectors are all discussed. Security guard responsibilities in case of fire are also outlined.

TYPICAL PROBLEMS

Access of Maintenance Employees to Buildings and Valuables

Experience repeatedly confirms that thefts may occur when individuals are permitted to work alone near money or valuables. Many retail stores and merchandise warehouses have an established policy that a lone employee cannot remain in the building after hours. This rule is based on safety considerations as well as security, since a single employee may not be able to obtain help if injured by machinery or equipment.

Yet some businesses allow repairmen, painters, floor waxers, maintenance workers, and janitorial employees to remain on the premises after hours. This is in spite of the fact that far less may be known about these workers than about the firm's regular employees.

It is undisputed that most janitorial employees are dependable, honest persons in every respect. But there are always exceptions. Sometimes people who seek janitorial work are unable to compete and are only marginally employable. This may relate to background problems involving frequent arrests, lack of skills, drinking or drug habits, or excessive absences.

To maintain control over janitorial or maintenance workers, management usually must either restrict work activities or provide supervision. Some firms obtain this control by scheduling janitorial duties while company supervisors and employees are still at work. But this practice may interfere with customers coming into the building, and there is always the possibility of a lawsuit if a customer trips on a freshly mopped floor.

Some firms have solved this security problem by giving the janitor a key to the office only, keeping stockrooms, warehouses, and money drawers securely locked. The janitor then works at night, without supervision.

This system was used by a Des Moines, Iowa, firm for a time, until shortages were found in the company warehouse adjoining the offices. Investigation eventually determined that the janitor regularly gained access to the warehouse and stole merchandise. He accomplished this by pushing upward on office ceiling tiles and gaining access to the ceiling crawl space that opened into the warehouse.

A Long Beach, California, company followed a similar procedure in giving an office key to the night janitor. Convinced that the firm's safe contained a large amount of cash, the janitor spent an entire weekend breaking open the safe with a heavy wrecking bar. Enraged when he found only $53 in the safe, the janitor set fire to the office. He then smashed the front door as he left, believing the entry would be blamed on an outside burglar.

A water sprinkler system put out the office fire, but only after sprinkler water had flowed into the warehouse and caused extensive damage. Eventually, after one of his fingerprints was found inside the battered safe, the janitor confessed and pleaded guilty in court.

In a warehouse or retail store, management sometimes locks janitors or floor waxers in the building, then checks these workers out when a supervisor returns to the building. But in some instances it may be illegal to lock personnel in a building unless some doors are equipped with emergency fire exits, and this renders the system ineffective unless the emergency exit is the kind that signals or records each instance of emergency departure. Obviously, locking janitorial workers in the building will not prevent theft if merchandise can be passed out through windows or other openings.

The policy of a Palo Alto, California, electronics company was to use one janitorial employee in a large warehouse at night. Doors to this facility were connected with an alarm system, so management would be made aware if the janitor opened one of the doors. There were no windows in this building. Carrying expensive electronic components out through an unsecured roof hatch, the janitor used a rope to lower these items to a confederate waiting outside the building.

Not content with these thefts, the unsupervised janitor wrapped other expensive merchandise in the manner used to send United Parcel (UPS) shipments. Using a label with a fictitious addressee on the wrapped packages, the thief mingled his packages with legitimate prewrapped shipments awaiting UPS pickup on the following day. The janitor was discovered when the firm's security officer compared the number of shipping tickets with the number of packages being picked up by the UPS delivery driver.

Tools and Equipment

A good security system utilizes controls over equipment, tools, and machinery. Whether some of these items are satisfactorily regulated or misused is frequently the difference between good security and serious business loss. While the security representative does not have direct responsibility for the maintenance of equipment and tools, there is a definite relation of interest.

Tool Thefts

Tool and equipment thefts are a continuing problem in security. Usually, a company is not aware of the extent of losses unless there is an inventory of tools and equipment on hand. It is therefore suggested that all tools and maintenance items be listed on a master inventory, and that this list be brought up to date on a yearly basis.

A charge-out system, used by all employees, will establish responsibility of individuals *temporarily* taking tools and equipment. This can be a very simple form, listing serial number, description of the tool, employee issuing, and the signature of the person who acknowledges receipt. Charge-outs should be dated and kept in file until the item is returned.

Experience shows that tool cribs, tool cabinets, and storage lockers should be locked at night, or at any time when supervision is not present. If employees need tools on a continuing basis, issuance may be considered on a permanent basis, rather than on a temporary charge-out.

Firms that do not have a large number of expensive tools sometimes maintain these on a plywood wall board, tracing the outline of individual tools in paint against the backboard. By glancing at the board, supervisors can tell at any given time whether tools are missing, especially at the end of a work shift. Immediate inquiries can be made to locate the missing items.

Marking the Firm's Tools

It is recommended that the company mark all tools with the firm name. There are three basic reasons behind this recommendation: (1) a thief will often hesitate to steal something if he/she realizes the article can be identified; (2) identification makes it possible for a police agency to locate the owner when stolen property is recovered; and (3) marking allows property to be positively identified as evidence in court during a criminal prosecution against the thief.

Various methods have been used to mark or identify tools. Some companies have merely painted the handles of company tools with a distinctive color of paint. If tool losses are considerable, it may be worthwhile to utilize an electric tool marking kit, etching the company name or initials on the handle of each. The hard tungsten-carbide tip of the electric tool marker makes it useful in etching on almost any material—metal, wood, plastic, or glass. Rather than write out the full name of the firm, some businesses etch out identification numbers or symbols.

Invisible fluids are also commercially available to paint company identification on valuable items. These marks are visible under a fluorescent light. This system has a drawback, however, since a potential thief may be deterred by a clearly marked tool.

Replacement Tools

Experience shows that employees frequently go to a company stockroom and request a new tool, claiming that they have broken the item being used. In those cases it is suggested that a replacement tool not be issued until the broken item is turned in to the stockroom inventory clerk.

Searching Employees for Tools

If hand tool losses are a continuing problem, security representatives may request management to consider installation of a metal detector near the employee exit. Portable, hand-held metal detectors are available at reasonable cost, but the effectiveness of these units is sometimes limited. Efficient metal detectors, of the types used to detect weapons carried by airline passengers, are available for industrial use.

If a search of departing employees is made for tools, the procedures used must be uniformly applied to all individuals in the line. Otherwise, a lawsuit charging discrimination can be sustained against the company.

It is also advisable to inform employees in writing at the time of hire that all departing workers may be subject to search. This should be stated as a precondition of employment, and the applicant requested to sign a written statement acknowledging this right on the part of the company.

Automobile Supplies and Gasoline

The security representative should consider controls for automobile tires, batteries, spark plugs, and other parts. A heavy-duty chain can be used to secure automobile tires, or a 1-inch steel pipe can be fabricated to provide a lockable control system, if it is not possible to maintain tires and parts in a locked storeroom.

Employees who are otherwise honest sometimes show no hesitancy in stealing company gasoline. At the minimum, the company should maintain a daily gasoline log, with entries made every time gasoline is issued. This log should record the number of the vehicle, the quantity of gasoline pumped, and the mileage on the vehicle odometer. A regular comparison of total vehicle usage with the quantity of gasoline purchased should reveal a difference of no more than approximately 2 percent for shrinkage.

The maintenance of a gasoline usage log is subject to employee failure to record all entries. Better control systems are available, with most requiring the insertion of a prenumbered ticket into the mechanism before the pump will be activated. When gasoline is issued, a system of this kind prints the exact amount onto the ticket. Tickets are controlled numerically, with an audit copy retained inside the pump mechanism for verification.

In a recent case in Baltimore, the company gasoline pump was located alongside a fence that marked the boundary of the employee parking lot. Employees took turns in passing the pump hose through the wire fence to personal automobiles that were filled in rotation. Installation of a short section of steel mesh behind the pump prevented a continuance of these employee thefts.

Dual controls for the gasoline pump should be considered, with a good padlock on the pump proper and an electrical shutoff inside a nearby building to discontinue electrical power after supervision has left for the day. The electrical switch should also be locked.

Owing to gasoline shortages in recent years, some firms have converted delivery trucks and forklifts to propane fuel. While this has been taking place, some employees themselves have been changing over to pickup trucks, campers, and passenger vehicles powered by this same fuel. Not realizing that propane is of value to workers, some companies have neglected to set up controls over propane fuel.

Welding Equipment

Welding equipment is often used in business and industrial applications. If burglars discover that a cutting torch is available on the premises, they may use this equipment to cut into the company safe.

Usually poorly trained in the use of welding equipment, burglars may start an accidental fire. In some states, the penalty for burglary with a cutting torch is so severe that professional burglars hesitate to use this kind of equipment. Nevertheless, the fire hazard is one of the security factors that may make it inadvisable to locate a computer room in the vicinity of a business office safe used for the protection of money. This is because fire is one of the most serious threats to a computer installation.

Welding equipment is often a desirable item for thieves, and it is therefore advisable to lock up welding torch heads and equipment when not in use. Pry bars and other heavy tools should also be locked up, if there is any possibility these devices could be used to open the company safe.

Pallets and Hand Trucks

Some companies retain reasonably good control over merchandise but neglect to protect items such as pallets and hand trucks. Wooden pallets, or skids, cost a great deal, especially if they are made of hardwood. Since it is often easier to load and unload by placing merchandise on pallets, there should be accountability when pallets are left on the customer's dock. The driver should bring back his own pallets or obtain pallets of equal value.

Since pallets are frequently stolen and sold if they cannot be identified, it is suggested the company name be painted on each pallet by use of a stencil and a spray gun. Workers sometimes steal pallets, since the hardwood content provides good fuel for residential fireplaces.

The Forklift as a Security Threat

Development of the forklift provided business and industry with a tool of great value for transporting and storing merchandise in bulk. Despite this usefulness, the forklift may figure in some serious problems in the areas of both security and safety.

If a burglar gets his hands on a forklift, he can load stolen goods without delay. Time is essential to the burglar, who will stay on the scene of the crime only long enough to enter and carry away the loot.

If a company safe is not too heavy, thieves may use the prongs of a forklift to rip out a warehouse or office door, forklift the strongbox, and load it into the bed of a pickup truck. Such loss may be prevented by controlling access to the company forklift.

Some companies have a firm rule requiring forklift keys to be removed and locked in a secure place at the close of business. Of course, there is still a possibility that a burglar may have the skills needed to "hot wire" the ignition on this piece of equipment, but this forces the criminal to spend more time in the building.

Recently, burglars tried a series of warehouse doors before they found one that could be forced open. The only merchandise in the warehouse of interest to the thieves was at the end of the warehouse opposite the forced door. There, small appliances worth $35,000 were stored on wooden pallets. The distance from the door would have made burglary impractical, expect for the fact that the key had been left in the company forklift. With the aid of this equipment, the thieves quickly loaded the merchandise and transported it to Baja California and Mexico, where it was sold on the black market.

In San Francisco, thieves broke into one warehouse solely to obtain the use of a forklift. Driving this equipment five blocks down the street after breaking through a warehouse door, the burglars reached the scene of their intended burglary. They battered through the wall of a second building, obtaining a large quantity of sporting goods that would not otherwise have been available.

In a case in Chicago, a burglar made entry into a warehouse by cutting a hole in the roof and dropping into the interior of a warehouse by sliding down a rope. Once inside, the burglar discovered that he was trapped, since he could not climb back up the rope attached to the roof. Using an available forklift, the intruder ripped out an expensive steel roll-down door and carried away merchandise from the warehouse stock.

In Philadelphia, a liquor wholesaler lost about $12,000 when a disgruntled forklift operator deliberately broke cases of expensive whiskey and French wines because the employee was not permitted to drink company beer with his lunch.

These examples make it apparent that forklift access should be controlled if good security is to be maintained.

Company Uniforms

Uniforms furnished by a company allow workers to save on clothing expenditures and make individuals readily identifiable with their jobs. The original cost of uniforms is usually absorbed by the company, with laundry or cleaning provided on a regular basis.

If uniforms are readily available, some employees assume that they have a right to help themselves to uniforms at any time. But uniform replacement costs may be considerable.

Investigation by the security department of a major airline recently revealed that employees stole flight jackets to avoid spending money for winter clothes. In warmer weather, employees carried away large quantities of coveralls for use in home gardening, fishing trips, and odd jobs.

Another investigation involving mechanics revealed that employees regularly discarded coveralls that became soiled, instead of turning in dirty uniforms when new ones were issued.

Experience shows that many employees do not regard this kind of appropriation as theft until company losses are made clear to them. Some firms have utilized security awareness sessions to point out individual responsibilities to workers. Garments should be issued on an individual basis, requiring the surrender of a soiled uniform before clean items can be furnished; the uniform supply should be retained in a locked storeroom.

Trash Disposal Problems

The company trash container is frequently used as a place of concealment by the company thief. Trash is seldom subject to inspection when taken out. If a thief hides merchandise in an outside dumpster or bin, it is usually easy to return after working hours and remove stolen items from concealment. And if merchandise should be detected in outgoing trash, the responsible employee can claim that it got there by mistake.

The theft problem here may be avoided if trash is taken out only at specific times, and then only under the supervision of a security man or responsible official. A procedure for collapsing boxes is also helpful, since space in the dumpster may be at a premium and concealed articles are likely to be exposed when cardboard cartons are collapsed and folded.

It is sometimes difficult to maintain good security if trash is taken out by janitorial personnel, without supervision. The likelihood for theft may be increased if trash is left to be taken out by janitorial workers on a night shift.

Some firms utilize a procedure in which the trash dumpster is rolled back inside the company warehouse at the close of the working day. This may effectively reduce theft, since concealed items are not available after closing time. But the presence of trash inside the building may be an increased fire hazard. Some managers feel the trash dumpster should not be brought back inside the building without written approval of the firm's fire insurance company.

Some companies have effectively controlled trash dumpster thefts by padlocking the dumpster after all trash has been put in the container. Frequently, this will not be permitted by the trash collection service, however, because time may be lost in locating the company supervisor holding the key and in removing the lock.

Use of a trash compactor should eliminate many of the problems associated with trash theft. Caution should be exercised in the use of this equipment, however, if there is a possibility that a neighborhood child could get into the machinery and be crushed. In addition to concern for the life of the child, management must be aware that any serious injury could result in a costly lawsuit.

Some firms feel there is less likelihood for concealment of company merchandise if clear plastic trash bags are used. Their assumption is that employees would hesitate before putting stolen items in a container in which the stolen merchandise could be easily seen.

Many financial institutions retain all trash in plastic sacks inside a locked room, disposing of trash on a daily basis about one week after collection. Trash is retained for a time in case checks or negotiable items have been thrown away which can be recovered by a subsequent search. In a disposal system of this kind, a thief can still use trash bags to conceal stolen items but must have access to the trash storage room in order to complete the theft.

Unsaleable or obsolete items are regularly thrown into the trash by some firms. Without this practice, there may not be enough shelf space in the warehouse for merchandise stocks. Smashing of this discarded merchandise at the time of disposal is to be recommended. Experience indicates that some individuals may remove discarded merchandise from the trash and present it for a refund from an earlier purchase, unless the junked item is smashed.

To make certain that merchandise deliberately put into the trash can be distinguished from items hidden by a thief, some firms remove the property from inventory by writing a "no charge" sales ticket. The item is then smashed.

Warehouses and businesses sometimes use a conveyor belt system for trash removal. This carries empty boxes out a wall opening, which sometimes offers easy building access to a burglar. Since trash boxes frequently remain on the endless belt, a vandal's match tossed into the outside trash bin or compactor could spread fire into the conveyor belt opening.

A tight-fitting, locked metal door may be needed to secure the conveyor opening during hours of nonoperation.

Stolen merchandise may not be the only item of interest in the company trash. A recent computer security survey for a Las Vegas, Nevada, bank disclosed that printout sheets containing customer account information sometimes blew out of the trash dumpster. In another incident of this kind, an Atlanta, Georgia, bank lost a costly lawsuit that resulted from its failure to control this type of customer information.

Confidential trash should obviously be made unreadable by running this material through a paper shredder. If there is any question, management should decide what company records can safely be thrown away, and the security director should advise management of any violations observed.

The manager of a food processing plant in Maryland was under the impression that placing confidential documents in the company paper shredder would automatically take care of the disposal problem. The cutting blades in some paper shredders can be set wide apart, and a curious employee at this plant demonstrated that shredded sheets could be reconstructed by an industrial spy with a minimum of work.

Security Personnel Doubling in Safety

Management has a responsibility to set up and implement company programs in both security and safety. While duties and responsibilities in these two fields are different, the goal of both safety and security is to reduce or eliminate company loss. Functions of the two departments can sometimes be joined.

Major firms often staff a security department and a safety department, with both answering to a director of loss prevention. Smaller firms may not feel justified in hiring both a full-time security director and a full-time safety director, so they combine the responsibilities in one official.

In any company, security employees may be asked to back up individuals responsible for safety. Fire is one of the major threats to life and property in any business. All security employees should therefore understand the basics of fire protection, as well as the systems used by their own firm.

A business is seldom able to avoid loss in the event of fire, regardless of the amount of fire insurance coverage that is in effect. If unable to deliver merchandise, a firm may lose customers and good will; and rebuilding is always expensive.

PROVIDING SECURITY THROUGH FIRE PROTECTION

Uninformed persons frequently believe that a fireman simply connects a section of hose, opens a hydrant, and pours water on a fire. The impression is that anyone with enough muscle can do the job. But fire fighting is a technical operation and requires specialized study and skills. Training is also helpful in preventing the outbreak of fires.

Fires start and burn by following natural scientific laws. These laws are dependable, exact, and orderly. When these are understood, fire hazards and causes will become apparent.

Fire results from the combination of (1) heat, (2) fuel, and (3) oxygen. When a substance that will burn is heated to a certain critical temperature called its "ignition temperature," it will ignite and continue to burn so long as there is fuel, heat, and a supply of oxygen. For example, when a heat source (lighted cigarette) and a fuel supply (discarded newspaper) come together under the proper conditions, we can predict that a fire will break out. This is the so-called fire triangle (Figure 26).

FIGURE 26   Fire triangle.

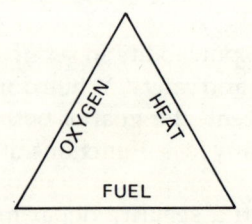

To stop a fire after it has started, one element in the fire triangle must be removed (see Figure 27). For example, if you turn a water hose on a wood fire, the heat will be removed; when you throw a blanket on a clothing fire, the oxygen will be removed; if you take unburned logs out of a fireplace, the fire will go out because additional fuel is not available.

FIGURE 27   Extinguishment triangle.

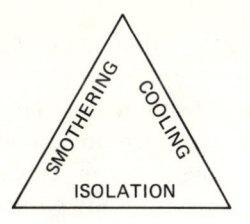

```
        SMOTHERING   COOLING
              ISOLATION
```

Water Sprinkler Systems

Water cools down wood and many other fuels to a temperature so low that they will not burn. Automatic water sprinkler systems are therefore very effective in putting out fires or in containing blazes until firefighters arrive. Fire insurance companies will sometimes grant reductions in the cost of fire insurance if a sprinkler system is installed in a business or warehouse. Over a period of years, premium savings may be more than enough to pay for the installation of a system of this kind.

These systems are automatic, since they will be activated when fire conditions are reached, whether or not individuals are in the building. In an installation of the kind commonly used, sprinkler heads are mounted in pipes in the ceiling of the building. Sprinkler heads are plugged with a fusible metal that will melt at a specific temperature, such as 165°. When the metal is melted by fire, the water sprays out through the open head. Sprinkler heads are most effective if they spray water over a wide area. In a fire prevention inspection, it should be pointed out that supplies or merchandise should never be stacked so that water spray is impeded. Locating sprinkler heads closer than 18 inches from the ceiling should not be permitted, since clearance is needed for an effective spray. Also, ceiling insulation that has worked loose should not be allowed to block the effectiveness of sprinkler heads.

There are two major types of automatic water sprinkler systems, the wet pipe system and the dry pipe system. In parts of the country where temperatures are very cold, a wet pipe system cannot be utilized. This is because a wet system already has water in the sprinkler lines, and the water in the installation would freeze in cold weather. In a dry pipe system, water for fighting the fire is kept in a reserve container in a part of the building that remains heated. When sprinkler heads are opened, water from the reserve container is sucked into the lines and sprays through the heads before the water reaches a freezing temperature.

An automatic sprinkler system is effective only on fires that break out below the sprinkler system. A fire resulting from a firebomb on the roof, or from other causes, may spread out of control before the sprinkler system begins to operate.

An outside post indicator valve is usually installed to control the water supply to an automatic sprinkler system, although hand wheels and other valves are used. Heavy losses have been sustained by businesses after an embittered employee turned off the control valve and set the building on fire during night hours. Under fire codes used in most areas, a business may use a heavy-grade padlock to secure a post indicator or wheel valve in an open position. A short length of heavy chain may be needed with the padlock in securing a wheel valve.

Although the possibility of sabotage may be reduced by locking the post indicator or wheel valve, it should be pointed out that serious damage can result to merchandise if the water supply is locked in the open position. Most water sprinkler systems can be monitored by a central alarm station, alerting the central station operator in the event the system begins to discharge, or in the event water pressures drop to a dangerous level.

At times a sprinkler head may discharge without warning when the fusible plug in the sprinkler head becomes loose, or a head may be broken off if struck by a forklift operator. To minimize water damage, some firms furnish a key to a security supervisor, chief engineer, or warehouse superintendent. If the post indicator valve or wheel valve is shut off to prevent water damage, some insurance policies require that an employee remain at the valve until repairs to the sprinkler system have been made. This is because a fire could break out inside the building, and the insurance company would prefer to pay for water damage rather than to have the entire building burn down.

Actual cases indicate that automatic sprinkler systems should be tested regularly, with water lines blown out to remove corrosion and sediment. There is always a possibility here that the engineer testing the water lines could fail to recharge the water lines, because of outside distractions. Some firms utilize a system in which the engineer has the basic responsibility to test and reactivate the system, with a follow-up verification by a security supervisor.

Classes of Fires

Authorities on fire prevention classify fires into three general types:

Class A—Fires in ordinary combustibles such as paper, packing boxes, wood, cloth, etc., which are normally extinguished by cooling down the burning fuel.

Class B—Fires involving flammable liquids such as gasoline, oil, naptha, etc., which are best extinguished by smothering.

Class C—Fires involving electrical installations, appliances, and wiring, in which the use of an extinguishing agent that is a nonconductor of electricity will prevent electrocution.

Use of Extinguishers or Portable Equipment

The smaller the fire, the easier it is to extinguish. Authorities on fire fighting report that in ordinary combustibles, fire can multiply itself 50 times in 8 minutes. In highly flammable substances such as paint, grease, or oil, the rate of increase may be even higher. Unless the fire is contained or extinguished from the outset, it may be out of control.

Authorities recommend that the fire department be notified at the outset, but portable extinguishing equipment should be used as soon as possible. Portable hand extinguishers, strategically placed, properly maintained and used, are indispensable in any fire protection program, even in locations where automatic water sprinkler systems and stand pipe and hose have been installed.

The three classes of fires are best extinguished by different methods. Some are put out by lowering the temperature of the burning fuel below its ignition point by cooling or quenching. Others are extinguished by cutting off the supply of oxygen through smothering or blanketing, and still others by a combination of cooling and smothering.

In a fire prevention plan, it is important to anticipate the type of fire that is likely to occur and to provide extinguishers for that specific class of fire.

Classes of Extinguishers

Extinguishers are classified as A, B, and C, according to the type of fire they are designed to put out. Water has the greatest cooling effect of any readily available substance. Since a cooling effect is needed for class A fires, water is the principal ingredient in all class A, or cooling type, extinguishers.

Class B extinguishers smother fires by cutting off the oxygen supply. Contents of class B extinguishers vary; there are a number of powdered chemicals and foam-producing substances, dense, heavier-than-air gases, and quick-vaporizing liquids that may be used.

Class C extinguishers use an extinguishing agent that does not conduct electricity.

Class A Type	Class B Type	Class C Type
Soda and acid Water tank with pressure or gas cartridge	Foam Carbon dioxide (CO_2) Dry chemical—pressure type	Carbon dioxide (CO_2) Dry chemical—pressure type

There are some extinguishers (A, B, C combination type) that may be satisfactory in fighting any class of fire. But there are reasons why this universal type of extinguisher is not always used. A class A type extinguisher costs considerably less to purchase and operate, and the class A extinguisher may be more effective in fighting the class of fire usually encountered (class A fire).

It is important to provide and use the correct type of extinguisher. In some cases, application of a class A extinguisher to a class B fire may actually spread flaming oil, gasoline, etc., over a wider area, causing more harm than good. Application of a class A extinguisher to an electrical fire can result in electric shock to the firefighter. Security personnel and other employees should be aware of these potential hazards.

Placement and Use of Hand Extinguishers

Security or safety inspections should be made regularly in business and industrial establishments to make certain that extinguishers are available, properly placed, and not blocked by merchandise. This inspection should verify that the right kinds of extinguishers have been put into service, and that they have been properly filled.

To make it easy for employees to find extinguishers in time of need, the location of each piece of equipment should be clearly marked to conform to OSHA standards.[1] Tags on extinguishers should clearly state when the equipment was filled and tested.

Surveys in some companies show that 80 to 90 percent of employees have no familiarity with hand extinguishers. Experience shows that in time of emergency many of these employees would be hesitant or slow to use the equipment. Affording training to all employees may be impractical, but a representative number on each shift should be given a basic indoctrination. In most cities, the local fire department will provide instruction and technical help without charge. Many companies that contract to refill hand extinguishers will allow employees to discharge the contents of extinguishers under supervision, giving employees familiarity with the equipment.

[1] Occupational Safety and Health Act. See Chapter 17.

Extinguishers that hold contents under pressure should be given hydrostatic tests. With the passage of time, the metal case of the extinguisher may corrode and split or leak, creating the possibility of injury to an employee.

Fire Hoses

Fire hoses are very helpful in combating fire; however, this equipment has limitations which should be pointed out. The average individual simply may not have the experience or muscle power to handle a standard fire hose. That is why business and commercial installations frequently install hose lines of smaller size. It should be realized that city firemen will probably not be able to couple standard equipment to company water connections or hoses.

If the water supply is not adequate, then the building fire hose may be of no value. A Reno, Nevada, security guard recently rushed to the top floor of an apartment building complex, where a fire had broken out. The guard passed by a hand extinguisher in his rush to reach the building fire hose, only to find that water pressure was not adequate at the top floor level.

Extreme cold is another condition under which a company fire hose may not operate.

Some companies that have relied on fire hoses have found the hoses unusable in an emergency because of rot. This problem may be expected unless the hose is of good grade, protected from the weather, and tested for reliability. Advice from fire equipment suppliers should be obtained if there is any question as to serviceability of equipment.

Fire Doors

Many buildings are constructed with brick or masonry fire walls that can separate the building into individual compartments. Even a serious fire can frequently be contained until professional firemen arrive if doors between the fire walls close properly.

Many fire doors are of metal or metal-covered construction, sliding on a wheel or pulley arrangement that will close automatically. Frequently, these doors are held in an open position by a rope or chain with a fusible metal link that will melt at a low temperature. When the fusible link melts, the door will be forced closed by weights that pull the door from the opposite direction. The door then forms part of the fire wall that separates the building into compartments.

Security guards at the building should be trained to close fire doors manually without waiting for the links to melt. This helps to eliminate drafts that spread the fire. Guards not actively fighting the blaze should close all doors and windows as employees evacuate, excepting doors or windows leading to fire escapes or exits.

If the guard is at an industrial plant with manually controlled sprinklers, the guard should be trained to use the system.

Employees in a warehouse or stockroom frequently pile merchandise against a fire door or block the door in other ways. A safety and security inspection of the building should result in immediate correction of conditions of this kind.

Fire Escapes

An intruder can sometimes use a fire escape to gain access to a building. If the escape has a swinging section stairs for the last flight down, a burglar may use a long hook to pull the section within reach. The escape may then be used as a walkway to get to the second floor. If this appears to be a problem in controlling building access, it is suggested that the door at the top of the fire escape be locked from the outside, with an emergency exit crash bar on the inside. It may also be helpful to place an alarm on this door.

If a fire escape has swinging section stairs for the last flight down, a building guard may be trained to lower the section at the sounding of the fire alarm, unless, of course, the guard is occupied with fighting a fire in that immediate area.

A company security representative may sometimes observe that a fire escape has been blocked by merchandise or damaged by a company delivery truck. It is suggested that immediate repairs be recommended to management, to make certain that employees are not injured in an emergency. Repairs should prevent injuries to employees as well as a serious lawsuit against the company.

Fire Detectors

A variety of fire detectors are available for business or home use. Some of these detect smoke, flame, a sudden increase in temperature, or atmospheric changes that result from fire.

Perhaps the most common detector in business or industrial applications is the so-called rate of rise detector. Activated by a sudden increase in temperature, this kind of detector is very reliable and reasonable in cost. In common with other detectors, a rate of rise sensor can be connected to an audible alarm, a central alarm station, or a printer at a guard house.

Since rate of rise detectors are activated by variations in temperature, they are seldom used in residential installations. Some detectors cost somewhat more but usually are very satisfactory in home applications.

Ionization type detectors are activated by changes in the structure of ions in the atmosphere and may be set off when products of combustion change the chemical content of gasses in the air. They are usually used in computer room applications or industrial areas where a very sensitive detector is desired. These detectors are regarded as too sensitive for many home or business applications.

Additional Guard Activity in Case of Fire

A guard who is not directly involved in fighting a fire can nevertheless be of great value in reducing fire damage if he is properly trained. A Miami, Florida, bank sustained serious fire loss because city firemen could not quickly find the location of the blaze in a multistory building. In case of fire, a guard or employee should have responsibility to go to the street and direct the fire department to the scene.

If building occupants are in danger, the guard should open exits and direct occupants through them in an orderly manner.

Losses may be seriously reduced if highly combustible building stocks near the blaze can be moved promptly. If tarpaulins are available, guards can be trained to throw them over computer equipment or very expensive merchandise to prevent water damage.

In a small industrial operation, machines may be operating although there is no one present except a guard. It may be advisable to instruct the guard to cut off machinery in the affected area. Power for lighting purposes usually should be left on unless a computer installation is involved or unless electricity may endanger the lives of employees or city firemen.

A guard can sometimes reduce loss by moving company delivery trucks, automobiles, or boxcars located near a fire.

SUMMARY

While maintenance and janitorial employees are usually reliable, security breaches are more likely to occur when employees are allowed inside a business after hours without supervision.

Tools should be charged out to responsible employees and marked as company property. It is also helpful to record serial numbers on equipment and tools on company inventory records.

Searches of employees for stolen tools should be made only if search procedures are applied uniformly to all departing employees. Management's right to search should be clearly stated as a precondition to accepting employment.

It is suggested replacement tools be issued only after return of broken or worn out items.

For good security, auto parts and supplies should be retained under lock and key, and gasoline issuance should be carefully supervised. Controls should extend to welding equipment, stockroom or warehouse pallets, and hand trucks.

Forklifts may be used by burglars or thieves, unless keys are controlled.

If clean company uniforms are issued only after soiled uniforms have been turned in, losses of uniforms will usually be minimized.

Historically, the trash container behind a store or business has been the location where stolen merchandise is most frequently concealed. Ideally, trash containers should be spot-checked by security or management and then locked, if possible. Discarded boxes should be folded flat when put into the trash container.

Fires never occur without the presence of three elements: (1) heat, (2) fuel, and (3) oxygen. Cutting off the availability of any one element will put out the fire.

The water sprinkler system is the most commonly used type of fire protection for business or institutional buildings. Two types of overhead water sprinkler systems are installed: (1) the wet type and (2) the dry type. Systems are activated when fire gets hot enough to melt a fusible link in a sprinkler head, allowing a spray of water to come through the head. Merchandise should never be stacked closer than 18 inches from the heads, so the water spray is unimpeded. Sprinkler systems should be tested regularly, and usually control valves should be locked in the open position to prevent sabotage.

There are three basic classes of fires: (1) class A—wood, paper, and ordinary combustibles, best extinguished by water; (2) class B—flammable liquids such as gasoline, oil, or naptha, best controlled by smothering; and (3) class C—electrical fires, for which a nonconductor of electricity is needed.

Employees should be trained to use extinguishers with a caution that the right type is necessary—an A, B, or C extinguisher for the corresponding type of blaze.

Ideally, fire doors should be alarmed for security, but care should be exercised to make sure they conform to local fire laws.

Guards away from the location of the fire should direct employees, close fire doors, and direct firemen to the scene of the blaze.

REVIEW AND DISCUSSION

1. Outline the problems that may result when maintenance employees, janitorial workers, or other employees are in the building after hours without supervision.
2. Describe some of the controls that can be used to maintain security. What are the objections to locking employees inside the building?
3. How may a charge-out system be used to control tools?
4. Give reasons why tools should be marked as company property.
5. Under what conditions may departing employees be searched for company tools?
6. Outline how a forklift may be used by criminals in burglary and theft situations.
7. How is the trash dumpster or bin used by company thieves? What steps can be taken to eliminate this problem?
8. What is the so-called fire triangle?
9. How does an automatic water sprinkler system work?
10. What is the difference between a wet pipe and a dry pipe sprinkler system?
11. What are the three classes of fires, class A, class B, and class C?
12. What types of extinguishers should be used on each class of fire?
13. Discuss requirements for the use of extinguishers by employees.
14. What are the potential benefits of fire hoses? What are potential failures related to fire hoses?
15. How do fire doors work?
16. Describe three types of fire detectors and how they work.
17. List ways in which a guard can help reduce fire loss even though he is not in the immediate vicinity of the fire.

5

Alarms and Other Protective Systems

This chapter is concerned with alarms—their uses and effectiveness, how different systems are used to alert the police, and how different sensors or triggering devices get the signal of an intrusion. Closed circuit television installations (CCTV) are discussed. The last part of the chapter focuses on guard forces, guard patrols, and other protective systems.

ALARMS

Simply put, an alarm is a device that gives a warning. It is an assembly of equipment arranged to signal the beginning or the presence of a hazard needing attention.

In the sense most commonly used in security, an alarm system is designed to alert someone to the presence of an intruder, such as a burglar or robber in a business, institutional property, or home. Such a system is frequently called an intrusion alarm.

Other systems may be used to warn of some predetermined undesirable or dangerous condition, such as a critical lowering of temperature in an industrial chemical process, a shortage of fuel in a generator, the outbreak of fire, or the discharge of a fire suppressant (carbon dioxide) in a computer installation that could asphyxiate persons remaining in the computer room. An alarm system of this type may also signal that a machine is running at too fast a rate. Almost any condition relating to temperature, humidity, water flow, electrical power usage, or industrial process can be monitored by an alarm system. This type of system may be called a physical condition alarm.

While physical condition alarms may not relate directly to security, they may require an immediate response by armed guards or other security personnel who are on the premises after engineers, maintenance employees, and company officials have left for the day.

Basic Parts of an Alarm System

Considerable training is needed to qualify as an expert in alarm systems. This is a special field, since there are many kinds of alarms and engineering applications in use. Alarms are almost constantly subject to change and improvement.

All alarm systems have three common elements, regardless of the individual design or components. These three parts are:

1. A sensor, or triggering device that senses the entry of an intruder or the existence of a physical condition or hazard that needs attention.
2. A circuit that is activated by the sensor, carrying a message to the signaling apparatus.
3. A signaling system or device, activated by the circuit. The signaling mechanism may sound a loud bell, flash a blinking red light, punch a message on a tape, dial a telephone call to the police, or warn in a combination of ways.

Benefits and Uses of an Intrusion Alarm

Installed and properly handled, an intrusion alarm system can be expected to set off the chain of events leading to catching the entering criminal before serious harm occurs. But there is an additional benefit that can be anticipated from such a system: It is a psychological deterrent, convincing most criminals to stay away.

Obviously, no burglar wants to be captured by police responding to an alarm. But there are times when the criminal enters in the mistaken belief that the building is not protected. Accordingly, it is usually advisable to let the criminal element know that the property is alarmed, by appropriate signs, by word of mouth, or by other means.

82   Alarms and Other Protective Systems

While of great value, alarm systems in themselves are not a complete answer to all security needs. They back up perimeter security, but they must be part of a coordinated program. Alarms keep no one out. They simply alert to the fact that action should be taken. It may be pointless to put an alarm on a warehouse door, for example, unless someone is available and has been trained to respond. In most installations, therefore, the alarm system must be designed in such a way that it evokes optimum response from the security department, available guards, or management representatives.

As far as cost is concerned, many facilities that now use several guards can save money by placing most entrances under an alarm system and retaining only enough guards to respond to intrusions and other emergency needs.

Assessing the Need for an Alarm System

When providing security recommendations, some individuals invariably advise every business and industrial establishment to install an alarm system. Although frequently justified, this advice may be an error in some instances.

In assessing the need for an alarm system, several questions should be asked: What are the anticipated loss factors? How likely is someone to get in? What success can the intruder anticipate in obtaining money or merchandise of value? How certain is an alarm system to prevent this loss? What are the costs of installing and maintaining an alarm system, as compared with the costs of anticipated losses?

It is to be noted that all merchandise may be desirable to thieves, but some types are difficult to haul away, or easy to trace, which makes them difficult to dispose of on the black market. The types of merchandise handled should therefore be considered; jewelry, for example, almost always merits alarm protection.

Another factor to be evaluated is whether the facility in question has had a history of losses from robbery, burglary, or theft. Is the building on a well-lighted street? In a good neighborhood? Effectively patrolled? Does it have good perimeter protection? Fencing? Good protection on doors and windows?

Different Kinds of Systems That Get the Word to Police or Management

There are three commonly used response systems that notify police or management that an alarm has been activated. These are:

1. *The central station.* This is a private clearing house set up to receive alarm signals from subscribers or customers who have alarm sensors installed on their premises. The central station is a control center, with personnel on duty continuously to investigate trouble alarms and to notify police or firemen when an intrusion or fire signal is received. Some central stations are owned by guard companies, who send their own uniformed representatives to the scene of the alarm.

   An intruder, of course, is not aware that a silent signal has been sent to the central station. Police authorities usually like this kind of system, since it enables officers to catch the criminal while the crime is taking place.

   The so-called *proprietary system* is a kind of central station, operated and owned by the same company or facility where the alarm sensors are installed. While police or firemen may be notified by an alarm condition, the response is usually made by the facility's own security or company fire department.

   There is another variation of the central station system. In some localities, alarm signals are received on a board in the local police station. This activity is being phased out in most law enforcement agencies, since most police departments no longer have the time to monitor the incoming signals.

2. *The local alarm system (bell-ringer alarm).* This is a low-cost system, installed when the expense of a reliable response from a central station system cannot be justified. Most installations of this kind activate a loudly ringing bell located on the wall above the front door of the premises. A siren, horn, or flashing red light may take the place of the more common ringing bell. The difference in cost between the two systems arises because a leased telephone line carries the signal to a central station, whereas a local alarm sounds only at the site.

   The ringing alarm will usually frighten away an intruder but is usually not very effective in helping law enforcement apprehend a criminal. Many authorities on crime believe a far more effective deterrent to crime is to catch the culprit red-handed.

   There is no certainty that the police will ever be notified when a local or bell-ringer alarm has been activated. The store or shop installing the system may intend that neighbors will call the police. But unless persons nearby actually see someone in the building, the local alarm may continue to ring unanswered for hours. Many individuals consider this type of alarm installation a neighborhood nuisance. Then, too, there may be no one in the neighborhood who can hear the bell when it rings during the night.

3. *Automatic telephone dialers.* An automatic telephone dialer, with a tape-recording attachment, sends a silent alarm signal. This system

can be set to dial predetermined numbers for both the police department and the business owner. When the system is activated, a recording states that a break-in is in progress at the location covered by the system. Up to three telephone calls may be made.

This system is low in cost, both to install and to maintain. But if the telephone line is cut, or police lines are temporaruly busy, no message will be sent. In addition, if someone working with the intruder continually rings the telephone line going into the premises, the automatic dialing system cannot call outside. This is in contrast to the central station system, which uses a private leased line.

Automatic telephone dialers are also somewhat prone to false alarms that tie up police telephone lines. It should be pointed out that when a robbery or burglary alarm is received at a police station, available units usually "roll" at high speed. The surest way to ruin police cooperation is to install equipment that sends false alarms. Some city governments are also assessing heavy penalties for false alarms that are answered by police or fire units.

ALARM SENSORING OR TRIGGERING SYSTEMS

Electromechanical Devices—
Tape or Foil and Contacts

Perhaps the most commonly used business alarm systems utilize metal tape, foil, or wire on windows and glass surfaces, and magnetic contacts on the doors. A continuous electrical current that will not harm anyone flows through the tape or metal foil on windows. If the tape or foil is broken, the current is interrupted and the alarm is activated. This, of course, is a "fail safe" system.

Magnetic contacts are frequently used on doors or openings. With the door in a closed position, the magnet is mounted on the door and the switch on the door frame, so the magnet is aligned directly under the switch. When the door is opened, the magnetic field is withdrawn, the circuit is interrupted, and the alarm is triggered.

Systems of this kind can sometimes be wired around by bridging the circuits, and the tape or foil may break or be accidentally scratched, sending a false signal. A false alarm may be set off by wind blowing strongly against a door, causing the magnetic contact to separate from the switch.

Photoelectric Alarm Installation

In a photoelectric alarm system, a light beam is directed from a sending device to a receiving cell. If someone walks through the beam, the alarm goes off. This kind of system makes use of invisible light, so a burglar or other intruder cannot see the beam. Mirrors may be used to turn the beam around a corner. Under favorable conditions, the beam may be effective as far as 500 feet, although fog, haze, the light source intensity, and the use of mirrors may all cut down the limit of effectiveness.

While the light beam used is invisible, a burglar may be able to determine the location of installations of this kind by observing the housing of the system under daylight conditions. Unless several parallel beams are used, the burglar may be able to step over the beam or crawl under when the system is activated at night. If mirrors are used, a minor dislocation will disturb the alignment, making the system inoperable.

If a burglar should be able to climb over the beam without detection, however, he may still find it difficult to move large amounts of merchandise or valuables out of the building without crossing the light pattern.

Motion Detection Alarms

Motion detection alarms are of two kinds, some operating with the use of radio frequency transmission and some with ultrasonic waves.

Radio frequency systems transmit waves throughout the protected area from a transmitting to a receiving antenna. When the receiving antenna is adjusted to a specific level of emission, a disturbance of the wave pattern will set off the alarm. Unless the walls of the building are shielded, radio waves may penetrate to the outside and respond to motion outside the building. The system may therefore be prone to false alarms, unless skillfully installed.

Ultrasonic motion detectors utilize an installation similar to that of radio frequency systems. The device fills a space with a pattern of ultrasonic waves. Disturbance of the wave pattern by a moving object is detected and will activate the alarm. This system is less prone to false alarms than the radio frequency motion detector, since ultrasonic waves do not penetrate building walls.

Sonic Alarm Systems

Sonic alarm systems make use of noise to activate a detection system. Their design is based on the idea that any intruder will make some noise, even while walking across the floor. The installation includes a receiver with a microphone in the protected space, connected to an alarm signal. A system of this kind should not be considered in locations where background noise levels may cover up the noise made by someone breaking into the protected area.

Vibration Detectors

Vibration detector alarm systems utilize contact microphones or vibration sensors attached to inner walls or other areas of the building to be protected. The sensitivity of the alarm unit is usually adjustable, to cut out outside vibrations. Specific items, such as a master painting, can be protected by attaching a vibration sensor to the frame of the picture. Any movement or touch should set off the alarm. In specific applications this system may be very helpful, since it is not prone to sending false alarms.

Pressure Alarms

Pressure alarms may be used in stairwells or halls, concealed under carpeting or mats. Since the pressure device is quite flat, the device and the wires leading to it are easily hidden. Weight on the pressure device activiates switches that set off an alarm.

Fence Alarms

Taut wires are sometimes used in fence alarms, with the taut wire woven through the length of the fence fabric. Changes in the tension on the taut wire set off an alarm.

Beam alarm systems are sometimes used on the inside of a fence line. Laser beam installations have proved to be very dependable in outside locations.

Capacitance Alarms or Proximity Alarms

Capacitance detection systems, or proximity alarms as they are sometimes called, are used to protect metal containers or filing cabinets, as well as metal safes. The specific item to be protected must be an ungrounded metal object. Protection is provided by setting up an electromagnetic field, through wiring up to oscillator circuits which are set in balance. The electromagnetic field may extend several feet around the safe or object being protected. If a foreign object enters this field, the circuits are disrupted and the alarm is activated. All metallic objects located close to the protected object should be well grounded. On the other hand, the protected safe or cabinet should be completely isolated from the ground by a block of insulating material.

CLOSED CIRCUIT TELEVISION (CCTV)

Closed circuit television (CCTV) has contributed significantly to security protection. Properly installed and maintained, CCTV allows a guard or security representative to monitor different activities in widely separated areas. In a large bank, for example, constant supervision can be given to a "teller's line" with 20 or more teller's cages, an outside drive-in teller's window, the night depository and safe deposit box areas, the front lobby, and a locked gate for armored car entry.

With TV monitors, a warehouse guard can observe loading and unloading on two large docks on opposite sides of the building, an employee parking lot, and remote exits through which employees could carry out stolen merchandise.

In still another application, one guard using a combination of CCTV and employee identification cards can physically control one entryway while monitoring other employee entrances and a parking lot.

Equipment for assuring security may be very expensive, depending on the coverage desired. But it may be considerably less expensive to use TV than to pay the salaries of additional guards.

If a video tape recorder or camera is added to the TV equipment, criminal behavior or misconduct not only may be observed but also recorded for eventual identification of suspects and for the prosecutor to use in court.

Attempting to save money, some business and financial firms have made use of dummy CCTV cameras. While the use of dummy equipment may have psychological value as a deterrent, it is difficult to keep the truth from either employees or outsiders. Installing dummy equipment may also create the impression that management is making only half-hearted efforts to protect employees and property.

Workers in some industries have actually smashed CCTV installations, objecting to being watched by "big brother." In other cases, workers have sprayed paint over the viewing lens of camera equipment. Acceptance has become general in recent years, however, and there is no question as to management's right to oversee business and employee activities.

Business managers and supervisors sometimes express the attitude that installation of TV equipment has relieved them of responsibility for employee theft and other security problems. At best, TV is a helpful and practical tool, but the basic responsibility continues.

Experience indicates that TV installations should never be made without careful planning and consultation with technical experts in this field. After an installation is made, it is sometimes found that monitoring results are not up to expectations. One of the most common mistakes is in installing a CCTV camera that is not matched to the available light level. Ordinary TV cameras simply will not work properly where there is not sufficient light. Sometimes this can be corrected by changing light fixtures in the area under observation. Another approach is to install a more sensitive camera, one with a silicon target videocon.

Even when the installed camera is satisfactory, it may be of very limited value to a guard assigned to scan a large area such as a parking lot or loading dock. Such areas require a "pan, tilt, and zoom" installation that will allow the camera to swing about and give a close-in look at any questionable activity.

A CCTV installation may be needed in a stairway to observe unauthorized individuals using the stairs for access to a computer room or other restricted area. Unless the camera is installed at the top of the stairs, the intruder may move beyond the view of the camera before identification can be made. To protect the camera from tampering, it is recommended that the stairwell installation have a housing enclosing the electrical connections and bolts, at sufficient height to be out of reach of potential intruders.

Experience has shown that TV reception gradually deteriorates to the point of uselessness when TV cameras and monitors are not regularly checked by an expert in this field. A regular review program must be set up and followed if the installation is to be effective.

It should be emphasized that results cannot be expected from TV equipment if monitoring is not both continuous and conscientious. Without constant surveillance and dedicated monitoring employees, the installation of CCTV equipment may not justify the costs involved.

PROTECTION SERVICES

Business or home protection firms usually provide one or more of the following services:

1. Guard services, usually making use of uniformed and/or armed personnel, working regular shifts on the premises. These guards may be furnished on a contact by an outside company, or they may be employees of the firm being protected.

2. Patrol services, making a predetermined number of rounds through the property, especially on weekends and at night.
3. Guard dogs.

The Security Guard

Security guards may furnish the most effective protection available, reacting immediately to any unusual threat to life or property. Often their very presence deters potential intruders and criminals; they are trained to detect any criminal activity and to respond accordingly. Guards prevent employee theft and misconduct and insure employees' compliance with management's controls and rules.

A good guard program also furnishes employees a sense of personal security, protects employee cars and parking lots, and reduces the likelihood of attacks against company employees and their families.

Guards almost invariably provide a head start in reporting fire and in controlling serious fire damage. They are unlikely to send false alarms, are helpful in preventing accidents and personal injury, and are trained to direct people from the building quickly while maintaining order.

In addition, the guard may be the best ambassador for visitor assistance and good will in the entire company organization.

Because guards are human and can exercise judgment, they can perform beyond the capabilities of mechanical security devices. At the same time, guards can be subject to human errors and weaknesses. Guards should therefore be regularly supervised, and sometimes disciplined, to make certain they perform as management's representatives at all times.

Responsibilities and Limitations of Guards

A security guard is neither a policeman nor a peace officer. The guard's basic responsibility is to provide security for the persons and property he or she is hired to protect. Policemen have more duties and powers, with a basic responsibility to the public in general, and accordingly, policemen are provided more training.

A security guard usually acts in a preventive role to a greater degree than does a police officer. Usually the guard should be highly visible; those who may be considering crime or damage are likely to be deterred by his presence. Trained to be keenly observant, the guard is most effective when preventing violations from occurring. When a serious matter does arise, the guard takes emergency action but also calls for police and management assistance.

Police officers have the responsibility to enforce all public laws. A security guard may enforce some public laws, but usually only on company property or as they involve the company, the violations having taken place within the guard's view or presence. Unlike a policeman's public responsibility, a guard's responsibility is to the company for enforcing company rules and regulations with regard to company employees.

Guard Arrests

With rare exceptions, a security guard has only the legal powers of an ordinary citizen in making an arrest. Two conditions must exist before such an arrest can be made by a guard:

1. A violation of law (an offense) must have been committed.
2. It must be clear that the individual being arrested has committed the violation.

Both conditions must exist to the personal knowledge of the guard. Otherwise, the guard and the guard's employer may be subject to a lawsuit for false arrest.

If a guard observed an individual pick up a rock and throw it through the window of a company building, the guard could make an arrest. This is because breaking the window is a violation of law, and the guard saw the offender actually commit the act. It would be a different situation, however, if the guard did not actually see the rock thrown. In a case where the guard heard the breaking and rushed to the scene, no arrest could be made even if a group pointed to one of their number and said, "He did it." It would be clear that an offense had been committed, but the guard would not have personal knowledge of which individual was responsibile. It would be up to the guard to call the police for an investigation.

If an arrest is made, the guard has the responsibility to inform the individual that he or she is under arrest, as well as to indicate the charge. For example, the guard should say, "You are under arrest for theft."

If the suspect resists arrest, a guard is allowed to use reasonable force to subdue that person. If the suspect complies willingly, no force is necessary. If force is applied, it should be just enough force to prevent the suspect from escaping before the police arrive. The use of handcuffs will be justified on individuals who have resisted, or on suspects whose actions indicate that they may refuse to submit.

A guard should understand that anytime someone has been detained, that individual is actually under arrest, whether or not words to that effect were actually spoken. If the suspect concludes that he or she must stay to answer questions, or is not free to walk away, an arrest of the suspect has been made. If the suspect understands that he or she can walk away, but stays to clear up the matter with the police, no arrest has taken place. But the mere statement, "You have got to stay until officers arrive," constitutes an arrest.

A guard making an arrest should obtain evidence to make the case hold up in court. In general, evidence may be anything that proves the crime was committed by the suspect. Evidence may be any object or statement that would cause a reasonable jury to conclude the suspect committed the crime in question.

If a burglar was observed breaking into a store building, evidence could include pry marks left on a door, store merchandise (loot) found on the suspect, heel marks in the mud, and the guard's eye-witness account of what was actually observed.

If a guard sees the suspect steal something, it is necessary to make an arrest, followed by a search to recover the evidence. If the suspect is able to get rid of the stolen item, the case may not hold up in court.

A search should never be made by a guard until after the suspect has been placed under arrest. There are two types of objects that the guard should look for in making a search: (1) weapons such as guns, knives, etc., and (2) stolen property that may be held as evidence.

After an arrest has been made by a guard, the police should be called without any delay of any kind. As a follow-up to the arrest made by the guard, the police are required by law to take the suspect into custody and take over the investigation, handling the matter with prosecuting authorities.

Providing Weapons to Guards

The use of weapons by security guards is quite controversial. Year after year, guard mishandling of guns has led to serious lawsuits and verdicts against both private guard companies and individual employers. On the other hand, to take guns away from guards can lead to loss of life or property to outside holdup men or other violent criminals.

Laws in many localities have made training and regular practice on the firearms range mandatory for guards carrying guns.

Many authorities on security point out that guns may not be necessary unless police response is slow or unless unusually large amounts of money or valuable merchandise must be protected. Issuance of guns is recommended only under unusual circumstances, and after individual evaluation of each guard entrusted with the responsibility of a weapon.

Contract Guards vs. Company Guards

Arguments are sometimes advanced as to the superiority of contract guards over company guards and vice versa.

Contract guards may be more likely to perform from an objective, outside viewpoint. They are usually less likely to become involved in company politics or close associations.

Then, too, a company guard force is a secondary business. For all of management's ability to handle the basic problems of the firm, it usually lacks experience in setting up and handling a guard program. The argument is made that this is best left to experienced professionals.

A private business is usually hard pressed to provide adequate guard training, and often this training is not on a professional basis. Proper manuals and training materials may not be available or maintained for the use of guards at all times. It is essential that adequate instructions be given concerning the preparation of guard reports, so that guard supervisors and company managers will have a detailed, clear account of the problems encountered and reported by guards.

Actually, a contract guard service may cost a company less than hiring its own. When company guards are hired, the firm must meet costs of hiring, training, retirement, life insurance, social security, bonding and insurance, workmen's compensation, uniforms, equipment, and supplies.

A contract guard service may also provide some indirect benefits. In time of emergency, such as a strike, the size of the guard force can be increased almost at will, merely by a telephone call to the contract guard force office. If the needs of the firm slack off, the size of a contract guard force can be immediately reduced, without the firm's having to discharge its own employees. Should an individual guard not be to the liking of management, this guard can be easily transferred to an assignment with another firm.

If a union problem should develop within the guard force, this is the responsibility of the outside guard contractor rather than the firm contracting the service.

On the other hand, managers of some private firms feel they have more discipline and more control if guards answer to them alone rather than to an outside guard supplier. Managers also sometimes feel that guard loyalty to the employer is essential and that loyalty is stronger in a direct employer-employee relationship.

Watchclock Systems

Watchclock systems provide supervision of guards on weekends and holidays, during extended closings, and at night.

Alarms and Other Protective Systems   93

A guard carrying a clock visits a series of stations in a predetermined order. At each station the guard inserts a recording key into the portable watchclock. See Figure 28. Each key has raised numerals of an individual character, identifying that station. The key registers the station number and time on a paper tape in the clock. Only supervisors have access to the recording dial and tape record inside the clock. Guard supervisors or members of management can regularly review the entries on the tape, to see whether the guard actually walked the specified rounds throughout buildings and grounds.

Insurance companies sometimes give reduced rates on fire coverage to companies that require regular guard patrols. This reduction is based on the assumption that orderly patrol routines should provide a head start in reporting and controlling fires on the property. Insurance companies usually require that the tape recordings from watchclocks be retained for review and verification by representatives of the insurance company.

FIGURE 28   Guard carrying a watchclock at a keystation.

Photo courtesy of Detex Corporation, Chicago, Illinois.

Patrol Services

Patrol services utilize motorized or foot patrolmen to periodically check office buildings, industrial plants, and private homes. The usual practice is for a patrol employee to make a specified number of rounds during night hours or when the building is closed. Patrol services operating in waterfront areas may utilize boats.

Generally, the patrolman leaves a small tag in the door of the business, to show that the check was actually made as required. Patrol services sometimes use clock stations (watchclocks) for verfication of their rounds.

Since patrol services leave the property between rounds, the protection furnished is not continuous. The most serious weakness of a patrol service is that a burglar or other intruder can accurately predict when a security check will be made. The patrolman usually has a prescribed route of subscribers, and there is little or no opportunity to vary inspection times. A potential criminal can hide in the surrounding area and observe patrol schedules. The wrongdoer can then break into the building immediately after a patrol inspection, making an escape before the time of the next round.

Guard Dogs

Properly handled, guard dogs are of great value for security patrols. They can quickly locate a burglar hiding inside a warehouse or business—an intruder who would not be detected by regular guard patrols. These animals also provide excellent protection for the handler.

Dogs are sometimes locked into warehouses, industrial installations, and store buildings at night. Unusually effective as deterrents, these animals must be carefully confined inside a building or fenced area. While very helpful if handled properly, dogs may cause serious problems if handled carelessly. Serious injury and a costly lawsuit could result if an animal trained to attack should escape from the area of confinement.

The effectiveness of a guard dog may be impaired if a potential intruder is able to feed the dog regularly, or if poison is administered to the animal. Both the dog and the animal's handler should receive special training, and the dog should be taught to accept food only from the handler.

Dogs may obstruct or even endanger firemen, police officers, or maintenance personnel needing to come on the property under emergency conditions.

SUMMARY

There are three basic parts to an alarm system: (1) a sensor, or triggering device, (2) a circuit or electric line that carries the message received by the sensor, and (3) an alerting or signaling device that sounds a bell, flashes a light, prints out a message, or otherwise alerts someone to the alarm condition.

Intrusion alarms back up the physical protection systems and let someone know that the physical barriers have been penetrated.

Not every location can justify the costs of an alarm system. There are many variables that must be assessed here—among others, the extent and strength of physical protection, the value and portability of items being protected, and whether police or guards should respond to an alarm.

Three different systems are ordinarily used to alert police or management that an alarm has been set off: (1) A central station system receives an alarm on a direct line from the alarm installation (usually a silent signal); the central station then calls police or private security officers to take action. A *proprietary system* is a type of central alarm station, in which a large company or institution handles its own alarm installations and central station functions, and dispatches private police or calls local authorities to supplement private police. (2) A *local alarm* system (bell-ringer alarm) sounds only on the premises; thus it may not obtain a police response. (3) An *automatic telephone dialer* system can be triggered to inform three different parties that a break-in is occurring; the system is not always reliable.

Perhaps the most widely used perimeter alarm makes use of foil or wire on windows and magnetic contacts on doors. A continuous electrical current flows through the wire, and opening the window or door disrupts the circuit and sets off the alarm, which is essentially a "fail safe" system. Photoelectric alarms, also widely used, utilize an invisible light beam that is directed into a receiving cell. The alarm is set off when someone walks through the beam. Motion detection alarms are activated by motion that interferes with radio frequency transmission or ultrasonic waves. Slight noise from an intruder will set off a sonic alarm system, while vibration detector systems use contact microphones or vibration sensors that are sensitive to touch or movement. Capacitance or proximity alarms provide protection by setting up a specific magnetic field around the object to be protected; a disturbance of the magnetic field will set off the system.

Closed circuit TV (CCTV) has contributed significantly to security protection. With a proper installation, proper supervision can be given to a tellers' line of 20 or more tellers' cages in a bank, to large loading docks,

96 Alarms and Other Protective Systems

to employee parking lots, and to other facilities. But CCTV may not be worth the cost unless the right camera is used, the lighting facilities and conditions are optimal, and a conscientious monitor is available at all times to use the system's capabilities.

Security protection services include uniformed guards, patrol services and guard dogs. Security guards offer the advantage of an early start in reporting and combating fires, prevent accidents, and regulate the flow of traffic. Guards keep out intruders, thieves, and burglars and reduce losses from internal theft. In addition, they enforce administrative rules and controls laid down by management.

Firms will have to weigh both the advantages and disadvantages of using guards contracted from an outside service company as opposed to hiring their own guard force. In general, guards on contract are less expensive and less likely to be influenced by company politics.

An armed guard may be more effective in serious situations than one who does not carry a weapon, but in general it is recommended that guards be unarmed; an improperly trained guard may kill an employee or a bystander.

Patrol services are considerably less expensive than the services of uniformed guards, but guard coverage is continuous whereas patrols offer periods of nonsurveillance when a criminal may operate successfully.

Guard dogs are very effective if properly controlled, but management must be aware that a serious lawsuit may result if a dog injures someone who has a right to be in the building or if the dog manages to get out of the yard or building where it is confined.

REVIEW AND DISCUSSION

1. Define an alarm. What are the three basic parts of an alarm system? What is the function of each?
2. List some of the benefits and uses of intrusion alarms.
3. Do all businesses and industrial installations need an alarm system? How would you evaluate this need?
4. Describe the workings of a central station monitoring or response system.
5. What are the basic differences between a central station system and a local alarm system? What are the weaknesses of the latter?
6. What is the difference between a proprietary alarm installation and a central station?

Alarms and Other Protective Systems 97

7. Describe how an automatic telephone dialer works. What are the possible trouble spots in this kind of system?
8. Describe the workings of foil and contact systems in protecting the perimeter of a building.
9. How does a photoelectric alarm installation work? How could this system be vulnerable to a burglar?
10. Explain how two types of motion detector alarms work.
11. Describe sonic alarms, pressure mats, and a capacitance alarm system.
12. Describe the factors that may affect the performance of a CCTV system.
13. What are some of the mistakes commonly made in installing and maintaining CCTV?
14. What are the basic differences between the functions and responsibilities of a security guard and a police officer?
15. Outline some of the major problems encountered when guards make arrests.
16. Should weapons be given to uniformed guards? Defend your position.
17. What are the benefits of contract guards over company guards? Explain the benefits to be expected when a company hires its own guards.
18. How does a watchclock system work?
19. What are the differences between regular guard services and a patrol service?
20. Describe the benefits of and objections to furnishing guard dogs to a business or industrial client.

6

Access Controls

This chapter is concerned with access controls—methods for regulating individuals' entry to and departure from the property or business being protected. Problems of parking lot security, night lighting, and traffic and vehicle control are examined. In addition, consideration is given to controls over visitor and employee parking and over vendors and employees of contractors who come on the premises. The essential need for employee badges, visitor passes, and package passes is examined. Finally, material is presented concerning access and intrusion controls with proprietary electronic building security systems.

THE IMPORTANCE AND USE OF ACCESS CONTROLS

A security system can be expected to protect neither monetary assets nor employees unless there are good controls over ingress to and egress from the property. Security guards, fencing, lighting, electronic equipment, and locking devices may all be made ineffective by a lack of access controls. At the same time, it is to be kept in mind that access should usually be granted only on the basis of operational necessity.

The Parking Lot

Parking lots are usually needed to accommodate employees and visitors and to facilitate the control of traffic into and out of a business or industrial installation. A firm without adequate parking areas may find that business suffers as a result.

The employee parking lot may be a real benefit to workers in any metropolitan area. Some workers may refuse job offers if parking is not readily available. Companies that provide parking facilities have a legal obligation to provide adequate security for such areas, and may have a moral obligation as well.

Firms that employ women for nighttime service must be especially aware of parking lot security. A large airline, for example, may require stewardesses to report to work during nighttime hours, and female computer operators and hospital nurses commonly work late shifts. Unless good security can be provided for the areas in which these women park, many of them may decline to work at the institution or business in question. In addition, a company may find itself on the losing end of an expensive lawsuit if a woman is attacked or raped as a result of inadequate parking lot protection.

In spite of the need to conserve energy, night lighting on parking lots should be maintained, as well as security guard patrols adequate for both general activity and specific coverage of employee arrival and departure at shift breaks.

Good parking lot security will also ensure protection of employee vehicles, as well as batteries, tires, stereo equipment, and accessories. Personal articles may be stolen from employee cars unless patrols are adequate. Medical doctors on the staff of a hospital should be counseled to take their medical bags with them or to lock the bag in the trunk of the car.

Experience shows that some thefts in parking lot areas can be attributed to family members or friends of an employee who wait in the company parking lot for an employee to go off shift.

Usually, the greater the distance the employee must walk to an automobile, the less apt that individual is to carry merchandise from the business or industrial plant to a parked vehicle. If the employee passes by a loading dock en route to the parking area, the likelihood of theft increases. Whenever parking spaces are located quite close to a dock or warehouse area, a masonry wall of adequate height or a chain-link fence should be used to restrict access.

Companies with a large number of employees have found that issuing a vehicle pass or sticker helps a guard determine whether the vehicle has a right to park in the company lot. A system of this kind allows security representatives to identify unauthorized vehicles.

One of the problems with a vehicle sticker system of this kind is that the sticker may not be removed when the employee's vehicle is sold or traded in. Some vehicle passes of this kind can be removed from the automobile's windshield only by scraping with a razor blade. This, of course, usually destroys the sticker and it is preferable to issue the type that can be removed easily and destroyed immediately.

Identification problems can arise if the pass is not removed or if it is destroyed prior to transfer of the automobile. If the sticker is left in place, an intruder may use it to gain access to the company parking lot. If employees are told that another sticker will not be issued until they return the original sticker or can prove it has been destroyed, there is not much likelihood for misuse of vehicle stickers.

The Patrol Vehicle on the Parking Lot

Security guards can control with far greater effectiveness if provided with a patrol vehicle. Specially built vehicles help reduce manpower costs while still providing the essential coverage needed for a large parking lot area. Some authorities estimate that a guard using a patrol vehicle can cover four times the area covered on foot. A vehicle such as the Cushman patrol car can squeeze through openings less than 50 inches wide, providing cost-effective transportation. In addition, the patrol vehicle can provide protection for guards in bad weather and offers better visibility because of lights built onto the vehicle.

It should be pointed out, however, that a Cushman vehicle is not able to make much speed in response to an emergency situation.

Night Lighting

Light is an important factor in a environment. Darkness allows the criminal to use the element of surprise, makes identification difficult, and supplies the cover through which the criminal may escape.

As early as the fourth century, main crossroads in the city of Jerusalem were lighted with wood fires at night, apparently to prevent street crimes and to protect business establishments. The Arabs used a similar night lighting system in the streets of Cordoba, Spain, in the tenth century.

In today's world, good security lighting, both inside and outside a building, is a deterrent to crime. All doors, windows, and openings into a building should have adequate overhead lighting, with vandal-proof coverings.

Because night lighting is so important, it is advisable that these lights be turned on and off by an automatic timer. If the responsibility for activating the night lighting system is left to the last departing employee, there is a possibility for error. Timing cycles on an automatic control should be regularly changed to follow seasonal variances in sunset and sunrise.

Replacing Fixtures and Lights

Experience shows that most commercial establishments should have a regular inspection and replacement program for night lighting. It should also be pointed out that management should be certain that night lighting does not intefere with the operational needs of the business. A warehouse manager who is constantly seeking additional storage space may stack large merchandise boxes directly in front of a field of light from a light fixture, rendering the night lighting quite ineffective. Packing cases, truck equipment, pallets or wooden skids, and uncleared trash dumps may lessen the effectiveness of night lighting, if improperly located.

It may be advisable to protect outside lights by wire mesh covers or guards. If in spite of this precaution breakage of light fixtures continues to be a problem, break-resistant fixtures can be obtained from a lighting wholesaler or supplier. More expensive than ordinary fixtures, these installations will usually withstand attacks from juvenile vandals.

In a security inspection of a building at night, it may be reasonable to assume a burglary is in progress if interior lights are not working. In a situation of this kind, it is unwise to enter the building without police coverage.

Some Principles of Protective Lighting

When possible, exterior lighting should illuminate the roof of a business building. Since burglars like to work in the dark, exterior fuse boxes should be padlocked. As a general rule, the entire property should be illuminated; nearby street lighting alone is usually not adequate.

From time to time, it may be found that some areas around a building are well illuminated whereas others are inadequately lighted. Some security officers have utilized a camera light meter to locate these dark spots, so existing light fixtures can be adjusted for more uniform coverage.

Specialists on security lighting sometimes stress the need for high brightness contrast between a potential intruder and the background.

This contrast can be brought about by a combination of adequate lighting and light colors on painted surfaces. Two approaches can be used in this regard:

1. Light the area and all structures and buildings within the general boundaries of the premises.
2. Light the boundaries and approaches. This should place glaring lights in the face of the approaching intruder, leaving security guards or company employees in comparative darkness.

Additional Lighting for Some Areas

Experience indicates that consideration should be given to special lighting in some areas, depending on individual business requirements. These areas should include:

1. Building entryways.
2. Fence gates.
3. Employee parking lots.
4. Scale houses.
5. Guard houses.
6. Cargo cribs.
7. Computer buildings.
8. Any areas that have been the scene of past entries by intruders.

Night Lighting Standards

Security lighting needs will vary from location to location. In case of doubt, technical assistance should be requested from a member of The Society of Illuminating Engineers. An unscientific but practical test is to examine an average newspaper under outside lighting conditions. If subheads of the newspaper can be read at a glance, the lighting is usually considered adequate.

For more exacting standards, The Society of Illuminating Engineers recommends that light intensities be measured at ground level, with foot candle[1] requirements as follows:

Vital buildings, structures, and other sensitive locations: 2.0 foot candles
Pedestrian and vehicle areas: 2.0 foot candles
Unattended outdoor parking lots: 1.0 foot candle

[1] A foot candle is a measurement of the illumination produced at a surface, all points of which are at a distance of 1 foot from a uniform point source of 1 candle.

The Guard Shack

A guard shack can provide a central command post and communications center for a vehicle parking area.

In many situations, the guard shack can be used as a place to store records, to issue employee passes, to secure weapons and guns issued to guards, and to control traffic.

As a general rule, guard shacks should be located at major traffic arteries coming into and going out of the parking lot. Visibility should always be a consideration in selecting the exact location for a structure of this kind.

In considering the use of a guard shack, the security department should consider the following items, at the minimum:

1. Portability of a guard shack is sometimes a basic consideration. The guard shack may be needed as a dressing facility and clothing storage place for guards.
2. The guard shack may be needed to protect guards against inclement weather and to provide a place to eat lunch.
3. The guard shack should be equipped with a telephone in practically all instances and often also a radio receiver and transmitter, permitting close communication between the company security office and guards working in the field.
4. It may be necessary to insulate the guard shack, heating it in the winter and perhaps providing cooling in the summer.
5. A guard shack may provide more effective working space if tinted glass is used in the windows, if the guard shack is large enough to contain a rest room, and if it contains a cash drawer that is lockable.

Obviously, not all these considerations will apply to every location. A major industrial firm in Mexico recently bought a guard shack that was equipped with bullet-resistant glass and armor-plated sides. This industrial firm is located across the street from a major university that was the scene of student riots and sniping.

Closed Circuit TV

Closed-circuit television coverage can be used very effectively in some exterior installations. Unless utilized selectively, however, television coverage may not be worth the cost.

Generally, coverage is far more effective if the CCTV camera has "pan, tilt, and zoom" capability to allow the viewer to sweep over broad areas. It should also be remembered that CCTV installations are usually ineffective unless there is good exterior lighting in the areas under observation.

It should also be pointed out that CCTV coverage is usually a waste of money unless personnel are available to monitor the TV screen adequately and to call for immediate assistance if a potential problem is observed. Additionally, TV monitoring and camera units need regular inspection and replacement if they are to remain effective.

BADGE SYSTEMS AS A MEANS OF CONTROL

Some companies hire so few employees that all are acquainted and are familiar with each other's job requirements. A badge system may not be needed in such a company. In larger firms, there is frequently a definite need for control by the use of badge systems.

Properly used, a badge system may keep unauthorized individuals off the premises, limit employee access to confidential areas, and serve to identify any individuals on the premises who may be lost.

Objections to a badge system are frequently expressed, especially by female employees. This is because the clip or pin used to hold the badge can harm clothing. Some employees also object to the use of a badge as a type of forced regimentation. On the other hand, some employees feel that the badge has psychological value, since it makes an employee feel "part of the team."

Experience shows that consideration should be given to including the following items on a company badge:

Employee photograph
Employee physical description (usually on back)
Employee signature
Employee social security number
Employee number
Date of issuance
Place of employee assignment (by symbol or color)
Authorizing signature

It may also be helpful to have a numerical sequence control on each badge, readily visible, as well as the badge's expiration date. Experienced security representatives also recommend that badges be produced and issued to employees as soon after hire as possible. If there is a delay of several days to one week in providing a badge, regular employees may waste time in questioning the new person.

If there are sensitive areas inside the installation, it is usually helpful to use colors or other coding devices such as numbers on employee badges in order to control internal access.

Whenever an identification badge is reported lost or stolen, identifying information should be disseminated to all guards who control points of ingress. This "hot sheet" system is essential to prevent unauthorized access and as a tool to retrieve identification badges if an illegal penetration should be attempted. This list of invalid badges should include badge numbers of terminated employees who did not turn in their badges.

The Temporary Badge

When an employee leaves an identification badge at home, a temporary badge should be issued, with a one-day expiration date imprinted thereon. This badge should be given to the employee only after the employee has been positively identified and cleared by supervision. There should also be full accountability for each temporary badge. The temporary badge should be turned in as soon as a permanent badge is issued or as soon as the employee recovers the regular identification badge.

Visitor Badges

Normally, a visitor will have an appointment or be granted an appointment prior to being allowed access. Most visitor badge systems require that the company receptionist in the lobby or the guard at the front station be notified prior to the time that a visitor is expected. The name of the visitor should be furnished, along with information as to which entrance of the building is most likely to be used and the time of the visitor's appointment.

Experience indicates that it is desirable to maintain a log at each entrance to the building, with spaces for the visitor's name, business connection, and the identity of the official visited. Notations should be included in this log as to the time of entry, time of departure, and badge number issued to the visitor. For control, a carbon copy of the badge issued by the receptionist or guard should be retained at a central guard station or collection point. Originals and carbons should be matched at the time of the visitor's departure. At the close of the shift, or at the close of the day, a check of those carbon copies still unmatched should be used to determine whether unaccounted-for visitors are still in the building. Upon arrival, the visitor should be announced to the company official by the receptionist or building guard. A visitor without an appointment should be required to remain in the reception area until approval is received. It is also best that a secretary or other employee escort the visitor both to and from an appointment.

Suggested form for a visitor pass is shown in Figure 29.

FIGURE 29 A two-part visitor pass form. When filled out by the visitor, it should be separated at the point of perforation. The upper half is then inserted into a badge holder and given to the visitor to wear; the lower half is retained by guard or receptionist.

```
               WESTERN WINE WAREHOUSE
                    VISITORS PASS

                         Expires:_____

  Name:_____
         Print - First & Last
  Person to visit_____

  Department_____

  Visitor's Company_____
- - - - - - - - - - - - - - - - - - - - - - -
  Purpose of visit_____
  _____

  _____
  Visitor's signature

  _____
  Escort's signature

  Date Issued:_____

  Building Entrance:_____

  Time in:_____

  Issuing guard or receptionist:_____
```

Suppliers and Vendors

Actual cases indicate that many thefts may be attributed to suppliers and delivery persons as well as vendors and outside building contractors whose access to the property is not controlled. Security in this area can be improved by utilizing only one gate or entryway to the property for the use of contractors and delivery persons. It is also desirable to issue vendor passes with numerical controls. If outside persons are required to wear these passes at all times, their identity will be apparent to company employees and security guards.

The receptionist or company security director should maintain a list of delivery persons and contractors who will be working on the premises.

This list should be posted at all security entrances. In controlling access, the security officer should have the following information on a card pertaining to specific suppliers' and contractors' representatives:

1. The supplier's or vendor's name, description, and company identification card number, the supplier's or vendor's telephone number, along with the signature of the authorizing representative of the outside firm.
2. The license number of the delivery vehicle.
3. The number of the badge issued to the delivery person, along with the date and place of issue.

It is usually helpful if the security officer prepares a daily card file or log record reflecting the destination of each entering delivery vehicle, the time of entry, the signature of the security officer granting access, and the invoice number of the freight bill or air bill involved in any package delivery. The contents of the package should be listed, if known, along with record of the number of individual boxes or packages to be delivered in the shipment.

If the delivery person should unload only part of the items on the incoming truck, this should be noted by the guard or receptionist when the truck departs.

A time clock may be used to show exactly when each vehicle arrived on the property and when it departed. Time on premises may be audited occasionally to determine whether a delivery person is spending an unreasonable amount of time on the company property. If this appears to be the case, investigation may reveal that the delivery person is operating a lottery or gambling game, or engaging in some other type of misconduct.

Nonemployees making deliveries should be given a copy in brief of company rules and regulations that pertain to exit and entry procedures and the control of delivery vehicles. Individuals making deliveries should be required to read this list of rules. Some firms have found it helpful to inform delivery drivers that they will be barred from future entry unless they conform completely with company rules.

It is suggested that permanent vendor passes not be issued unless the driver comes to the premises on a regular schedule and can be expected to continue on a regular basis. The vendor's supervisor should be told that a permanent vendor pass must be returned if the driver is transferred from the route or terminates employment. If the vendor understands that the delivery vehicle may not be allowed on the property unless company rules are followed, passes will usually be returned and security controls maintained.

Package Passes

Package passes should be utilized in many businesses. In the systems usually used, a guard or receptionist is on duty to check the packages being carried out by individuals leaving the premises. The package pass is usually retained or filed by the guard or receptionist who inspects the package.

It can be expected that thieves may prowl the halls of an office building, claiming to be repairmen while stealing property such as typewriters, business machines, or supplies. Experience also shows that employees will carry off all types of merchandise unless controls are used.

Whatever package pass system is used should apply to the tenants occupying rented space as well as to the owners of the remainder of the building.

Many firms use a package pass that has no carbon copies. With a system of this kind, there is no way that the guard can detect whether employees have added additional items to those for which approval was obtained. When a two-part form is used, one copy will remain in the approving supervisor's book. A security representative should then make a physical comparison of the supervisor's copy with the original package pass surrendered to the guard at the exit gate. A comparison of the supervisor's copy with the original copy will immediately reveal whether items have been added to the list. To forestall employee misconduct, it is suggested that management announce that security representatives will inspect contents of packages and will contact supervisory officials if there is a variation between the pass and the package contents, or if there is a variation between the original package pass and the copy retained in the supervisor's book.

Experience shows that the guard inspecting the outgoing package frequently cannot tell whether the signature on the pass is genuine. Security will be improved if the signatures of all authorized supervisors are placed on a single sheet of paper, retained under a protective cover at the guardhouse. With these signatures available, the guard can quickly compare the signature on the package pass with that of the approving supervisor.

It is suggested that signatures of the approving officials be written out on the package pass in full. Initials alone are much easier to forge. Signatures should also be written in ink rather than in pencil.

In practice, it has been found that supervisors are frequently lax in giving a specific description of the items that are covered by the package pass. Supervisors often give approval to remove items such as "a package of bologna," "a box of hand tools," or the like. It is suggested that the approving supervisor run a line through any spaces that are left blank on the package pass. This will help to keep on employee from adding unauthorized items.

When possible to do so, it is suggested that the package pass list the serial numbers on tools, typewriters, and machinery to be carried out.

Occasionally, a firm will use a package pass form as a control for lending company tools to employees. Of course, many firms have a strict policy against loaning company tools. But if the tool is one that has indefinite life and that an employee cannot afford to purchase, some companies allow the employee to use the tool at home.

If the package pass is used as a control for lending tools, it is suggested that an additional copy of the package pass be prepared at the time of issuance. This additional copy can be used as a "tickler" for review by supervision, to remind the employee to return the tool after a specified length of time.

Card-Key Access Systems

There are many ways to control access to property. Some firms use a security guard who grants entry to the premises on visual recognition alone. This type of system may prove inadequate for several reasons. During peak traffic periods the guard may make a mistaken identification or may not realize that the person seeking admission has been discharged a few days earlier. The guard may also be unable to recall who has been granted entry on a particular shift, even after the lapse of only a few hours. A system that does not rely on the memory of the guard uses a company badge that has also been manufactured to serve as a card-key for insertion into a door lock or entry terminal.

Since it is desirable to keep out persons who have been discharged, a card-key access system will reject badges or cards used as keys by terminated employees. A potential weakness of the card-key system is that an unauthorized person can steal or borrow a valid card that has been issued to someone else. This difficulty can be overcome, however, if the guard on duty is required to inspect the laminated photograph on an identification card of this kind. If the photograph matches the appearance of the card holder, the guard can allow the card to be inserted in the card reader, which in turn will open the access door if the card is valid.

A system of this kind can eliminate the need to issue keys to external doors. In addition, the need for manpower to unlock or lock remote doors is done away with. This system permits doors to be operated from great distances and offers good control over any number of doors or entryways.

The individual requesting admittance at a door or gate a long distance away can be identified at a central guard station using CCTV and the technique of comparing the photograph on the badge with the appearance of the individual observed on the TV monitoring screen. If the identification is in order, the employee then uses the badge as a card-key to gain entry.

Card-keys may also be used for a cost-effective alternative to the traditional clerical methods of time and attendance control. The entire operation can be done through a terminal which not only accepts employee badges as keys but also is connected to a system for recording time in and time out for each employee, by badge number.

Under some control systems, a so-called mantrap installation is used. This is a double-door entry booth, with access granted to the first door by a card-key lock activated by the employee badge. When the cardholder enters the booth, the first door locks shut behind the person seeking entry. When the door has been secured, a guard at a remote location instructs the cardholder to place the card-key on a reader. The guard can then observe the cardholder as well as the photo on the card, since the photo is projected on a reader screen located alongside a CCTV monitor that pictures the upper body of the cardholder. When identification is confirmed, the guard pushes a button releasing an electric strike on the inner door lock, allowing the cardholder to come through the second door. If the card-key is rejected, the person seeking entry is trapped inside the booth and must remain there until released by the guard.

Coupled with a minicomputer with memory capacity, a card-key control system can make a record of each identification card used to lock, unlock, or give status identification at each door. This record of when a specific card was used for entry and exit can be very helpful in a security investigation to determine which employees were on the premises or in a particular area at a given time. In a recent case in the Los Angeles area, an employee came into the building on visual recognition from a security guard. The employee then set fire in materials that smoldered for a considerable period of time before breaking out in flames. Later, an employee, possibly the same one, set another fire in the plant under similar circumstances. In neither case were the guards on duty able to state the identity of employees who had been granted access. In the investigation which followed these fires, police officials were unable to determine accurately which company employees had come onto the premises at the pertinent times. This was because the system allowed employees to enter on the visual recognition of guards, who were unable to recall specific employees who had passed through the gate.

As pointed out, the basic job of the card-key control system is the "people handling" aspect of the system that permits or limits access without confusion. But more sophisticated systems of this kind may also add an alarm-monitoring aspect to protect employees and property. The management of the firm, working with a security consultant, can select burglary alarms, fire alarms, panic button systems, and closed-circuit television devices from a complete line of alarm equipment. In addition, the functioning of vital machinery in the business activities or processes can be monitored. For example, the central control unit can be programmed to print out an alarm warning and to set off an audible alarm to alert the guard in the event monitoring devices detected the outbreak of fire or a machine failure in a frozen food room. By integrating controls, the card-key access system can be combined for both access control and alarm and equipment monitoring. A lone security guard stationed at the access control panel of this monitoring and control system can handle the functions formerly performed by a large number of guards. Controls can also be set up to record the quantity of gasoline or fuel pumped into a company truck, and the identity of the driver responsible for taking the fuel.

Company Tours

Some businesses permit visitors to tour their facilities. This is usually done as a good-will or advertising promotion. Some firms arrange regularly scheduled tours for student groups and organizations, while other firms will permit random visitors to walk in and join tours that are scheduled to go through the building. Obviously, there are some business benefits from these tours but there may also be drawbacks. Unless visitors are closely supervised, they may drift away from the main group and engage in theft or other misconduct.

Some burglars have used tours to gain access to offices or industrial facilities. Dropping out of the tour group, the burglar in this kind of situation hides out in a rest room or other place of concealment, waiting until after closing hours to burglarize the business.

As a matter of basic policy, security representatives usually recommend that tours not be taken through computer areas. A number of incidents are on file describing serious damage caused by outside groups taken on tours in computer areas. If tours are allowed, it is recommended that these visitors be individually registered and that the tour leader account for all individuals in the group at the end of the tour. Each visitor on a tour can be provided with a distinctive badge upon registration, and company employees can be informed that individuals wearing badges of that color are members of a tour. Badges used for tours should be accounted for numerically.

OTHER WEAKNESSES IN ACCESS

The Catering Truck

Through regular contact, the driver of a catering truck usually becomes well acquainted with employees at any business where the catering truck makes scheduled stops. Through familiarity, security rules are usually relaxed at break time. It is not unusual for a catering truck driver to wander about the building or to commit thefts. In conspiracy with inside employees, the driver may use the truck to carry out stolen merchandise. There is always the possibility, too, that the catering truck driver may work as an industrial spy for a competing firm.

Cases are also on file of employees who have traded company merchandise to the driver of a catering truck for food and drink items. Rules governing catering trucks should be set up and strictly adhered to. Drivers should be advised as to where trucks can be parked and where drivers can and cannot go. Spot checks should be made from time to time to ascertain whether employees give merchandise to the driver in exchange for beverages or food.

Employee Lockers

In some industries, employees are not allowed to work in street clothing. In a meat packing plant, for example, workers must change into a uniform, leaving personal clothing items in a dressing room locker.

Frequently, lockers are used as places to conceal stolen merchandise or money. Because of this, management should cultivate the idea that employee lockers are a privilege rather than a right.

Management may retain the legal authority to search company-owned lockers, if management clearly states this right as a condition of employment at the time of hire. The search procedure should be uniformly applied, so that a claim of discrimination cannot be made on the basis of an inspection of selected individuals in the work force.

From a security standpoint, lockers should not be installed near merchandise that is subject to theft. In addition, management should retain a master key to all lockers and individual employees should not be allowed to place their own padlocks on lockers.

Restricting Access Inside a Building or Facility

In examining security controls in multistory buildings, it is sometimes observed that high security areas have been designated on almost every floor. Usually, this dispersion is the result of management's using whatever space happened to be available at a particular time. Although access controls can be designed to protect each of these separate restricted areas, an ideal situation would be to relocate and consolidate all high security or sensitive areas on upper floors of the building. This would leave the several floors below as a buffer zone. Experience shows that it is usually easier to protect higher levels in a building than those at or near the ground.

SUMMARY

Perimeter controls, alarms, and other protective techniques may be ineffective unless access controls are properly set up and maintained. Access to the sales floor of a retail store, for example, must be granted to the general public, but access to that facility's stockroom or warehouse should be closely restricted. Job necessity for access should be the general restriction applied to all security operations.

Unless parking lots are properly lighted and patrolled, customers, employees, and visitors on the lot may be subject to robbery, assault or rape, and automobiles may be pilfered. Experience shows that the probability of theft increases when employee parking is permitted too close to freight docks or stockroom doors.

In industrial and business applications, special patrol vehicles have proved inexpensive and practical, greatly extending the range of security personnel.

Energy shortages have exerted pressure to curtail security lighting, but lighting continues to be one of the most effective security measures that can be used. To make certain lights remain operational, fixtures should be protected and a replacement and inspection program should be continuous. In general, outdoor lighting should be strong enough to provide a contrast between an intruder and a painted background. Obviously, some high-risk security areas may need extra security lighting. Technical assistance may be obtained from illuminating engineering specialists in large cities, but as a rough test, parking lot lighting is usually adequate if newspaper subheads can be read at a glance. Conditions vary, but in general 1.0 foot candle of light is adequate on an open parking lot, whereas 2.0 foot candles of light are needed in pedestrian traffic areas and vital buildings and entryways.

In selecting and locating a guard shack, consideration should be given to portability, visibility, storage, use of communications, heating in cold weather, and other specialized conditions of use.

An employee badge is often a vital tool in controlling access to the premises as well as in restricting access to areas in the interior. Temporary badges can be used to identify new workers or those who have left their badges at home.

A badge system for visitors is also helpful in controlling outsiders, especially if a two-part form is used and one part is retained as a control by the receptionist or guard. Experience shows that thefts may be reduced by instituting badge and entry control systems for suppliers and vendors who come on the premises.

Similarly, an effective package pass system regulates what can be taken off the property. A control copy, for comparison with the copy taken up by the exit guard, is necessary to make certain the pass holder does not carry out more than authorized.

Card-key access systems eliminate the need for guard to supervise all remote entrances in person. A card-key system will be far more effective if coupled to a minicomputer with a memory section to record when an individual card was used for access. Systems of this kind can be used to rule out an invalid card. Utilized with a CCTV camera, a guard can observe the person entering at a remote entryway, to make sure the valid card has not been stolen or loaned to someone else.

Department employees can be legally searched for stolen property if (1) all departing employees without exception are searched and (2) the right to search was specified as a precondition to employment at the time employment was accepted.

REVIEW AND DISCUSSION

1. What can be done to prevent thefts by persons in the parking lot who are waiting to pick up employees?
2. List the security precautions that may be taken to prevent assaults, vandalism, and theft on the parking lot.
3. What is the value of issuing stickers to use on employee cars for the parking lot? How may stickers be controlled?
4. What is a good, practical test to determine whether night lighting on a company or institutional parking lot is adequate?
5. List some special locations or areas that may require special night lighting.
6. What are the advantages and requirements of a guard shack?
7. How are badges used to control access? What should be on a badge? Why is it important for badges to be issued almost immediately after an employee goes to work?

8. What are the values and security hazards in using temporary badges?
9. Describe how a visitor badge program should be set up.
10. Why should vendors and suppliers coming on the property be controlled?
11. Outline the requirements of a good package pass system.
12. Describe how a card-key system, hooked to a minicomputer, can be used to monitor security, safety, and engineering functions of a building or a business complex.
13. What are some of the potential security risks in allowing a catering truck to come on the premises?
14. Under what conditions may the owner or manager of a business control employee lockers?
15. When and under what circumstances may a departing employee be searched?

7

Theft

This chapter examines the security problems caused by theft, exploring the real extent of theft in the United States and the consequences of theft to business. Also considered are whether most thefts involve the taking of money or merchandise, whether most losses are from external or internal thefts, and why employees steal. Some controls that may be set up are suggested. Finally, some common business theft situations are described: thefts in employee purchasing, thefts at the will-counter, the "early employee," driver collection thefts, and outside losses through till-tapping.

SOME BACKGROUND TO THE PROBLEM

What Is the Extent of Theft in the United States?

There is no certain answer as to how much is stolen each year. We can be sure only that theft losses are staggering. And this is a burden that most businesses must absorb or shift to the consumer.

Figures on theft losses are compiled regularly by the U.S. Department of Commerce, the Small Business Administration, bonding companies, and insurance groups. Estimates by two insurance companies set 1977 losses at about $7 billion. These figures are based on loss claims filed with insurance and bonding companies, but many firms are not insured, and losses suffered by these companies may never be recorded. Then, too, some merchandise thefts are simply never discovered, while pilferage incidents may be regarded as too minor to report.

Undoubtedly, the available statistics on theft are carefully compiled, yet in all likelihood they are incomplete. These statistics do, however, point out the seriousness of the problem.

What Are the Real Consequences of Theft in the Business World?

The full consequences of theft to a business may not be immediately apparent. Of course, the cash value of the goods or money taken can be calculated. But if merchandise is taken, a firm may not be able to fill a customer's order. The customer is then likely to turn to a competitor and may never return. Or if the items stolen are essential in a manufacturing operation, then the whole production line may be shut down.

It is also worth noting that insurance costs go up when theft claims are filed.

Employee morale may also suffer as a result of internal thefts, as some self-respecting employees simply do not want to work in an atmosphere where dishonesty is a way of life. Then, too, one theft may encourage another.

Do Thefts Involve Taking of Money or Merchandise?

Money is usually the most desirable item to the average theft. But large amounts of cash simply may not be available for the taking in many businesses. As a consequence, thieves take any kind of company property or goods that can be used or resold.

Exact figures are not available. But for many years insurance statisticians and security experts have been in general agreement that merchandise theft losses in business have exceeded cash losses by roughly seven to one. It is therefore apparent that a security program must be directed toward protecting merchandise and company property as well as money.

Sometimes security representatives for a business are lulled into a false sense of security. They begin to think that some kinds of merchandise are not subject to theft. But if merchandise is useful to a customer, then it must also be worth stealing.

Are Most Thefts from Inside or Outside the Business?

Both internal and external thieves cause serious security losses to both businesses and individuals. Most delivery truck drivers are conscientious, hard-working individuals. But sometimes a driver may steal from the incoming shipment, even before the delivery arrives at the business. Other thieves may grab unprotected freight before it can be brought in from the dock. And somewhat later, shoplifters and outside thieves may cause serious losses in store stocks and merchandise inventories.

But thefts by employees inside the business may be far more costly than thefts from outside. In 1977, statistical experts for three major insurance carriers stated that, for a number of years, dollar losses from internal theft were on average about four or five times greater then losses from external theft. There are exceptions to this, of course; a dress shop with honest clerks may sustain losses that are attributable only to shoplifters. In general, however, most losses to business through theft involve the taking of company merchandise by company employees.

Owners or managers of a business sometimes express considerable surprise when a long-time employee is discovered stealing. They sometimes lose sight of the fact that an individual at any level in a business may be involved. The experienced, trusted employee usually has the best opportunity to steal.

This does not mean that employees should not be trusted. But management and security employees should not relax those controls that are designed to prevent such problems.

Business managers sometimes regard theft as one of the costs of doing business, taking the attitude that theft can be neither completely eliminated nor controlled. But experience shows that an effective security representative can work to dispel this attitude. Uncontrolled theft, even on a small scale, may become a destructive force. A business cannot afford to set a level of theft above which the security department will begin action or a dollar value below which the theft act will be overlooked.

Why Employees Steal

No one really knows why employees steal. In many businesses employee pay scales are low. In retail stores and warehouses, company merchandise may be extremely attractive, and goods sometimes appear to be there for the taking.

Most workers regard themselves as honest at the time of hire. If an employee's attitude toward honesty changes, management may often be responsible. Sometimes the worker has a real or imaginary grievance against the company; in short, the employee feels he or she is owed something by the company. At times, management may give the impression that honesty really does not matter, especially when the company has no firm policy of maintaining integrity with regard to customers.

In a large number of cases of employee theft, there are three conditions that exist.

1. The thief believes that he or she has been wronged, and that taking something may settle the score. A grievance of this kind may, of course, be unjustified or imaginary.
2. The thief believes the taking of money or property may help to reach living standards or goals which have been set in the thiefs own mind. This may be an ego need, as conditions in this country are such that individuals are seldom compelled to steal to prevent starvation.
3. The opportunity for theft is comparatively easy, since security controls are inadequate or lacking altogether. This third condition is one that can often be eliminated by an effective security program. Where controls are adequate, dishonesty is usually at a minimum.

Some Security and Management Controls to Minimize Employee Theft

There are no infallible techniques to prevent employee theft. To be effective, procedures must be varied from time to time, from situation to situation, and from business to business.

The ways in which employees may steal are so varied that it would be difficult to study them all. In general, however, most firms have been successful in controlling employee theft by using two basic approaches:

1. Hiring honest and dependable people at the outset, keeping them oriented through good security and management attitudes, and reminding them of their security responsibilities. This is discussed in detail in the chapter on personnel security.
2. Using good merchandise and money controls, with unscheduled security inspections and audits to see that controls are enforced.

In setting up controls and audits over merchandise and money, two basic ideas are generally followed:

1. Reducing employee access to both money and merchandise, to the greatest extent practical. This means employees should have access strictly on the basis of need and for only as much time as necessary. At

other times, money and valuables should be kept under lock and key, with keys properly limited.
2. Separating responsibility between employees. The handling of money or merchandise should be entrusted to one employee, with independent accountability and record keeping for those items under the responsibility of a second employee.

The separation of responsibility idea is often used by accountants to prevent embezzlement, and it is frequently applied in security programs designed to prevent theft. Based on the premise that most employees are honest and can be expected to remain that way, separation of responsibility effectively eliminates even the temptation to steal. It is unlikely that two honest employees will become involved in theft at the same time, and any employee may hesitate to become involved in wrongdoing if he or she knows there is an independent verification of his activities.

An Example of an Audit

Recently, a retail store in the Midwest became lax in inspecting and auditing security controls at cash registers on the sales floor. Realizing that controls were being neglected, a male sales clerk closed out his cash register about an hour before the close of the store. The clerk then reopened the cash register with a new tape and rang up sales until closing time. He then removed the money rung up on the second tape, pocketing it. Since there was never an audit by supervisors or security representatives, the dishonest clerk was able to reinsert the original tape, time after time. Eventually, this series of thefts was detected when an audit showed that there was no continuity of numbers on audit tapes being used at this register.

In businesses where cash should be accounted for, it is essential to make regular audits of cash registers, without prior notification. The individual making the audit should take a register reading, subtract the opening reading and the starting cash, and compare this figure to the cash on hand (which is counted in the sales clerk's presence). Voids and refunds are then subtracted.

If the register does not balance, the sales clerk either has made an honest error or is trying to steal from the cash register. Repeated errors may indicate that the clerical employee simply does not have the ability to handle money or make change, rather than that he or she is dishonest.

If there is too much money in the cash register, the sales clerk may have been "underringing sales," waiting for an opportune time to remove the excess cash from the register when no one is looking.

If the register is found to be short, this may also be indicative of theft.

Some confirmed thieves make a practice of taking out a large bill when no one is observing, then underringing later transactions to bring the register back into balance.

If a register is found to be out of balance, questions should be asked in an attempt to resolve the variance. If differences are found in subsequent audits, close attention by supervisory and security personnel may be warranted.

Experience shows that there is no specific time of day when audits should be conducted. Days when supervisors are tied up with receiving or shipping problems and immediately before break periods or lunch hours are perhaps the most likely times to detect a theft.

KINDS OF THEFT BY EMPLOYEES

Employee Appropriation as a Form of Theft

Both management and security personnel should make it clear that employees are not free to use company merchandise or products without permission. It is not unusual to observe employees consuming lunch meat in a meat packing plant or cheese at a cheese distribution center.

The recently hired security director for an electronics warehouse reported that some employees failed to understand why they could not help themselves to stock parts to make a digital clock for home use. A security investigation at a wholesale bakery supply company revealed that workers made a habit of breaking open 70-pound boxes of shelled pecans, eating the nut meats while they went about their warehouse duties.

To prevent uncontrolled employee appropriation of the company product, a national manufacturer of razor blades allowed each employee in the factory to take enough razor blades for personal usage. Shortages were found to be excessive, however, and investigation disclosed that employees had begun to supply blades to large numbers of friends and relatives. Some workers at the factory went so far as to sell blades by the case at neighborhood bars. Management was eventually forced to stop furnishing blades to anyone, and to adopt a policy of prosecution for theft of any quantity of blades.

A major chicken processing plant in the South operated a feed mill and furnished feed and chickens to growers. Some spillage of the feed was unavoidable when it was loaded at the mill. For a number of years, management allowed mill employees to scoop up or sweep up this spilled feed and take the salvage home to feed their own farm animals. After a time, mill workers began deliberately to spill large quantities in loading. A short time later, the company security director caught employees taking the company feed in truckload lots.

Because of this tendency among some employees to "help themselves" many companies have a firm rule that any appropriation of company merchandise will be regarded as theft. Security employees then assist in the enforcement of this policy.

Taking Merchandise Samples

Successful company sales representatives often leave merchandise samples with prospective customers. Although giving away free samples can be an effective sales technique, uncontrolled taking of samples can be very costly to the manufacturer and may actually become a form of theft. Some firms do not allow salesmen in stockrooms or warehouses; instead, the salesmen must fill out a "sample request form" and submit it in advance of sales calls. This procedure allows management to evaluate the requests and to see that samples are taken only to proven customers or to likely prospects.

Theft of Side Products, Scrap, and Salvage

Many firms maintain good security controls over regular merchandise but are not so careful in protecting side products and salvage items. At times, the value of accumulations may not be realized.

An industrial uniform or linen supply company often accumulates scraps that are saleable as industrial rags, while an electrical wholesaler may stockpile sizeable lots of scrap copper. Any machine shop or metalworking operation may produce large amounts of scrap brass, aluminum, or stainless steel. X-ray plates contain silver; an X-ray technician in a hospital can sell patient X-rays for the silver salvage in the plates.

Frequently, management is unaware that an employee can steal these items without being branded as a thief. Often this is because security controls are lacking. Unless security and management officials are alert, warehousemen or production employees may sell these scrap accumulations out the back door of the business. And even if the sale is made in the interests of the business, the buyer may cheat the seller unless there is accountability by weight or count.

Often employees may throw away large amounts of saleable materials unless scrap bins are provided. Once bins are available, the security problem may be to keep employees from stealing from scrap containers.

If the business has a guard at its gate, it may be advisable for this security representative to inspect scrap shipments being taken from the premises. In a recent case in Cincinnati, Ohio, a search by a gate guard revealed that a large refuse bin, supposedly containing scrap iron, had a large quantity of scrap copper concealed under the iron. Investigation showed that the purchaser had paid the company for scrap iron but had managed to load scrap copper, which was worth considerably more.

A supermarket or grocery chain may accumulate large amounts of saleable scrap meat, bone, and fat. If these items are picked up on the premises, it is recommended that a security representative or supervisor observe each transaction, with a witness signing each weight ticket. Ideally, payments for scrap or salvage should be made by check, delivered to the business office of the company.

A form for control of scrap material or salvage sales is set out in Appendix 2. In using a control form, an accounting or management employee should compare all weight tickets and payments made by salvage firms to make certain that neither employees nor salvage companies are able to steal.

Experience shows that it may be helpful to place a "one-way" cover on scrap barrels. This cover permits easy insertion of materials but makes it difficult to remove items from inside. Employees sometimes stockpile items such as brass and copper scrap at their residences, removing small quantities from the scrap barrel on a regular basis.

In a recent case, the supervisor for a New Jersey firm placed chalk weight figures on steel barrels containing scrap. Through chance, the supervisor observed that an outside scrap buyer had brought an eraser and a piece of chalk, substituting his own weight figures to get more scrap than he actually paid for.

Employee Purchases

Most firms in the United States allow employees to purchase company merchandise, some giving employees a substantial discount. Transactions of this kind almost always represent a substantial benefit to rank and file workers. Some managers and owners of businesses feel employees are less likely to steal if granted this discount privilege.

Other companies, however, have reported unusual security problems arising from employee purchasing. Managers sometimes point to low profit margins where discounts are allowed. Some employees take undue advantage of the privilege, buying items for relatives and acquaintances who might have purchased them at regular prices. Then, too, some employees take advantage of their employer by placing extra items in their own package or by purchasing inexpensive items and then substituting more expensive goods after they have paid.

As a matter of security policy, all sales to employees should be handled by an approved manager or supervisor. In addition, the company should require that the selection, wrapping, or shipping of employee purchases be handled by an employee other than the purchaser.

It is also recommended that employees not be allowed to pick up purchases until they actually depart at the end of the work day. Some firms do not permit employees to return to the interior of the building after picking up purchases. Obviously, the pickup point should be as near to the exit door as possible.

Some firms have found that the management or supervisor handling employee sales did not actually ring up those sales individually and place the money in a cash register. Instead, the person was pocketing the proceeds at the end of the sale. Other companies have discovered that employees allowed access to collection areas or bookkeeping offices have pulled out and destroyed charge slips for employee purchases.

Dishonesty may also come to light if employees are allowed to buy defective merchandise or "seconds." Sometimes employees deliberately damage or scratch merchandise that they want to purchase at a reduced price. Because of experiences of this kind, some firms no longer allow employees to purchase company merchandise.

Theft at the Will Call Counter

Some retail and wholesale establishments provide added customer service by maintaining a will call counter. Theft is sometimes a serious problem at the will call counter, since it may be easy for the counter clerk to be in collusion with a dishonest customer.

In a recent case, a wholesale hardware company in Chicago found that a customer ordered only a single micrometer, but the will call clerk had placed an additional dozen micrometers into the customer's sack. Investigation disclosed that overdeliveries of this kind were being made to the customer on an almost daily basis. Selling the extra micrometers on the black market, the customer made regular payoffs to the dishonest counterman.

The manager of the hardware firm had earlier realized that there were significant stock discrepancies in warehouse merchandise but had shrugged these off as errors by the receiving and shipping clerks, or mistakes in bookkeeping.

Eventually, this collusion between the company employee and the customer was observed by a second employee who had been assigned to the will call counter. When confronted by the company security representative, the guilty clerk admitted passing out too much merchandise but claimed that he had merely made a human error because of the large number of customer orders being handled.

To prevent collusive theft of this kind, a member of a security representative can make unannounced audits of will call orders that are awaiting pickup. This kind of inspection or audit is based on the idea that the merchandise is selected by the will call clerk before the customer calls for the merchandise. Most companies follow a practice of receiving will call orders by telephone and immediately packaging the order.

For this kind of auditing to be effective, a package copy (floor copy) of the written sales order should be retained with the packaged merchandise on the will call shelves. If an inspection reveals that the will call clerk has set aside more merchandise than called for in the sales order (sales ticket), it does not necessarily mean that the sales clerk is dishonest. Anyone can make an error, especially an employee working under time pressure and selecting large numbers of orders. If a will call clerk consistently selects more merchandise than the order calls for, however, this may be a good indication that the will call clerk is dishonest.

Experience shows that it may be desirable to counsel the responsible employee if a selecting error is discovered. An investigation by the security representative is usually in order if a pattern appears to be developing in overselection of merchandise. Underselection can also damage customer relationships, as the customer may feel cheated. Errors of this kind are also costly to correct.

Even if will call orders are properly selected, there is always the possibility that the will call clerk may go back into the warehouse and obtain more merchandise at the time the customer comes to pick up the waiting package. This type of dishonest activity may be noticeable to other clerks or supervisors who are assigned to that area of the business.

The "Early Employee"

Employees report to work early for a variety of reasons, and in many instances this may be a commendable habit. At the same time, security representatives may keep in mind that some individuals arrive early because it is usually easier to steal when no one else is around.

Some firms that use time clocks do not allow employees to come in the building and punch in their time cards earlier than 15 minutes before work activities are to begin on that shift. One of the reasons for this time restriction is to allow supervisors to arrive at interior locations by the time rank and file workers report for assignments.

Merchandise Refunds

When merchandise is returned to a store for a refund, the transaction should be approved at the time by a supervisor. There are a number of reasons for this policy. In the first place, the so-called customer may have stolen the merchandise in question, either at this store or at another one in the area. Some thieves make a habit of picking up discarded sales tickets from store aisles and shoplifting an item of identical price that can be returned for cash.

Then, too, unless the supervisor actually verifies the return while the customer is still in the store, the transaction may represent a theft by the sales clerk, who has already removed money from the cash register in an amount that corresponds to the ticket. In this situation, there was no actual return of merchandise, although the sales clerk reported a return.

If the security representative is convinced that sales clerks have prepared fictitious refund slips by attempting to disguise handwriting or by raising the amounts on legitimate refunds, additional investigation may be warranted. A study may reveal that some sales clerks are involved in a great number of refund transactions while others are seldom involved. When refunds are frequent, examination of individual transactions by the security representative is usually justified.

An excessive number of void sales or overrings on a cash register may also be indicative of theft. This should not be taken as conclusive, however, as the sales clerk may not understand the cash register or the systems being used.

Driver Collection Losses

The money that route drivers collect for merchandise deliveries may be lost through theft, embezzlement, or armed robbery. Many companies have found it is unwise to allow drivers to go more than a day or two without turning in collections. If drivers are not regularly accountable some may begin to borrow from company funds. These employees may have every intention of replacing the company's money, but they may find themselves "in too deep" before the company realizes what has happened.

It is not unusual for a driver's collections to be stolen after reaching the company office. A janitor for an industrial linen supply company in San Francisco recently fished out drivers' receipt bags that had been dropped into a slot cut out of the top of a steel filing cabinet. After this experience, the firm purchased a commercial safe with a drop chute, built to resist tampering and such fishing expeditions.

Route drivers are also frequent victims of armed robbery, especially in slum neighborhoods in larger cities. Losses of this kind can usually be limited by dropping large bills into a routeman's safe, about 4 by 6 by 6 inches in size, bolted or welded to the metal floor or steel struts inside the truck. The driver should not be furnished a key to this safe, and a sign should be posted in the truck or on both sides, advising:

DRIVER DOES NOT HAVE KEY & CANNOT OPEN SAFE

Unless this precaution is taken, an armed robber may seriously injure the driver, assuming the driver can be beaten or threatened to open the strongbox.

Experience also shows that drivers may steal from each other unless check-in is carefully supervised and collections are locked up at the end of the working day.

Ticket Switching and Merchandise Substitution

Customers sometimes cheat retail stores by ticket or merchandise-switching techniques. In the usual case, the buyer waits until no one is watching and then transfers the ticket from a low-cost item to a high-cost item, disposing of the higher priced ticket. This kind of cheating can frequently be dealt with through teaching cashiers to stay alert to the going price of merchandise.

A number of types of self-destruct tickets are now available to stores experiencing this kind of problem. One ticket of this type is made so that it will break into shreds too fine to read when pulled off the merchandise. Other tickets can only be removed with a special tool retained by the cashier.

A helpful technique a store can use to minimize its losses is to double-mark items that are especially sensitive to price switching, placing an inked impression on the item itself and a tag or mark on the box that contains the item.

Some retail stores make a practice of placing the price stamp on the top or cap of bottled goods. A dishonest customer can effect a price switch merely by removing the cap and replacing it with another cap showing a lower price. This may be prevented by stamping the price on the merchandise proper, rather than on the cap.

Customers also participate in other types of fraud that may prove costly if not kept to a minimum. In a grocery store or supermarket, for example, a shopper may dump out the contents of an inexpensive box of prepared cereal, refilling the box with expensive cans of caviar or meat. Some very elaborate attempts of this kind are on record; in one, the thief even went so far as to reglue the top of the cereal box. Other attempts are quite crude, and accordingly more obvious.

In the final analysis, alert clerical employees, intent on good customer service, are usually helpful in keeping ticket switching or merchandise concealment at a low level in retail establishments.

Till-Tapping

One form of theft from a merchant is the till tap. The thief's problem here is to get a hand in the money drawer (till) or cash register without being observed. Till tappers usually operate in teams of two or three, with one member of the team as the driver of a getaway car outside the store. A second member of the team plays "Honest John," distracting the cashier's attention away from the money drawer. The third member, the tapper, then gets into the drawer, grabs available bills, and flees to the waiting car. Honest John and the tapper go into the store separately, of course, giving no indication that they are acquainted.

If a store follows good cash-handling practices and ensures that the cash register is closed after each transaction, there should be little possibility for a successful till tapping. In so far as possible, the cashier should remain close to the register, but in any case the cash register bell should alert the responsible employee to the opening of the drawer. Cashiers should be taught to wait on only one customer at a time and never to leave the register unattended without locking the register drawer and removing the key.

Employees should be briefed on this type of theft so that if a till tap occurs, they can recover in time to follow and identify the tapper and obtain the license number of the getaway car.

Some types of modern cash registers can be opened only if someone is standing on a rubber pressure pad immediately in front of the register. This usually eliminates the possibility that a criminal may reach over the counter and enter the register when a clerk is not nearby.

It is also wise to bolt the cash register to the counter, unless the register is very heavy. Till tappers have been known to grab the entire register and run to a waiting car rather than take the time to scoop currency from the drawer.

SUMMARY

No one knows the real cost of theft in the United States. We can be sure only that the losses are staggering. And the full consequences of theft in a business may not be immediately apparent. If key items are stolen, production lines may be shut down, or retail customers may go to a competitor if merchandise cannot be produced. Insurance costs rise and employee morale suffers in the event of internal theft. In most businesses, merchandise thefts exceed cash losses by about seven to one, and most losses are internal rather than external by about four or five to one. No one really knows why employees steal. Although a number of factors may be involved in theft, opportunity is the one ingredient that often can be eliminated or controlled by a good security program.

Most of the successful techniques to control internal losses are based on (1) hiring honest employees and cultivating and encouraging these individuals and (2) using good merchandise and money accountability systems, with unannounced audits or inspections to see that these controls are followed.

In setting up controls, two basic principles have been found workable: (1) reduce the number of employees who have access to valuables and reduce the time of access as much as practicable and (2) make one employee responsible for handling merchandise or money and a second responsible for record keeping and accountability.

Employees should not be allowed to use up or appropriate company products or merchandise or carry away merchandise samples without controls. Side products, scrap, and salvage may be as vulnerable to theft as company merchandise or cash. If employees are allowed to make purchases, transactions should be carefully monitored to make certain the employees do not take advantage of this privilege. The "early employee" may have come to work early only to commit a theft. Will-call transactions, merchandise refunds, and route driver collections should be audited and protected. Ticket-switching and till-tapping techniques can be brought to the attention of employees, who can then be alert to their use.

REVIEW AND DISCUSSION

1. How extensive is theft in the United States? Explain why loss figures may not be complete.
2. Is the value of merchandise stolen greater than the cost of money losses? What is the ratio between money and merchandise losses? Can you give some explanation for your answer?
3. Are external losses more expensive than internal losses? Here again, give the approximate ratio of losses and the factors that cause this difference.
4. Give some explanations as to why employees steal.

5. Outline and explain the two basic approaches management can take to establish security controls in order to eliminate or limit employee theft.
6. How may security inspections and controls be set up? Explain your statements.
7. What is meant by separation of responsibility between two employees? How may this idea be used? Give an example from a typical business situation.
8. How serious is employee pilferage or appropriation? Is this really a kind of theft? What should the security representative do about this kind of activity? Why?
9. Should there be controls over merchandise samples? Justify your answer.
10. Set out some ideas for the control of scrap, salvage, or side products in a business.
11. Explain how employees making purchases from their employer may cheat the business. How may this kind of dishonesty be controlled?
12. How can you control theft at the will call counter? Go into detail.
13. Is an "early employee" more likely to steal than a worker who arrives just in time to go to work? Why?
14. Explain how thefts may occur in merchandise refunding procedures.
15. Describe the basic security problems related to route driver collections. How may they be minimized?
16. Explain ticket-switching techniques and prevention methods.
17. Describe till-tapping and how it can be circumvented.

8

Shoplifting

The purpose of this chapter is to examine shoplifting with regard to who shoplifts, for what, and when as well as how merchandise arrangements, protective sales techniques, and judicious use of company employees and equipment can be used to control shoplifting. Legal aspects of shoplifting and how the merchant and the security department may avoid problems arising from making false arrests are also discussed.

INTRODUCTION

Shoplifting is a costly form of theft or larceny. Competition among businesses has reduced profit markups so greatly that there is often little margin left to cover customer theft. Preventing shoplifting losses may therefore make the difference between successful retail operation and a business that is on the verge of bankruptcy.

Who Shoplifts, for What, and When

A number of reliable studies have been made as to the type of person responsible for shoplifting. Results of these studies vary from location to location, and according to the type of store involved.

Undoubtedly, some professional shoplifters are responsible for part of this activity. Generally, these are criminals who do not engage in other types of criminal activity. Professionals steal on order, or steal the type of merchandise that is readily saleable on the black market. At times, several individuals may work together as professional shoplifters, with a scout in front and a guard in back to obstruct or slow up pursuit.

Aside from professionals, students are often involved in shoplifting, sometimes as members of a club that requires a successful theft as part of the initiation procedure. Some students pilfer as a team sport, much in the way that students drag race on a dare or play "chicken" to test the nerve of other students. In general, shoplifting is unusually heavy in stores located close to public schools. Studies show that housewives also are frequently responsible for shoplifting. In fact, on the whole, most shoplifting is perpetrated by students or housewives. It should be pointed out, however, that persons of both sexes, of all ages, and of all economic groups are involved in this type of activity.

Available studies show that shoplifters will steal almost any kind of merchandise. Ten months after opening, a large New York City motel reported the loss of 18,000 bath towels, 355 silver coffee pots, 38,000 demitasse spoons, 15,000 finger bowls, and 100 copies of the Bible.

Most items stolen by shoplifters are small and easy to conceal; some, however, are of considerable size. Cosmetics, earrings, lipsticks, perfumes, cigarettes, razor blades, scarves, phonograph tapes and records, and other jewelry items are among the merchandise frequently stolen.

In general, shoplifters prefer the crowded first floor of a store, large sales, and counters piled high with merchandise. They also prefer to take merchandise displayed close to exits. Professional shoplifters have occasionally admitted to planning their activity around employee shift changes, or at the time regular employees go to lunch or dinner.

Designing the Building to Reduce Shoplifting

If it appears likely that shoplifting may be a problem in a business building, the building should be designed so that all exits can be controlled. It is usually helpful to keep the number of exits to a minimum.

Experience shows that an L-shaped store lobby is difficult to keep under observation. Blind spots should be eliminated from the plan of a building, insofar as possible. Large pillars may also obstruct observation in some store layouts.

Some stores are designed with the office and accounting areas on a balcony, from which employees can look out over the sales floor. The possibility that employees on the balcony are watching activities on the sales floor tends to discourage shoplifters.

In arranging the store, it is usually helpful to locate large display boards on the perimeter of the sales floor or room. If displays are placed in the middle of the floor, shoplifters may conceal their activities behind these fixtures.

Within recent years, some retail stores, especially ladies' dress shops, have installed acrylic fixtures made of abrasion-resistant sheets. These see-through fixtures enable sales people to keep better tabs on goods displayed near entrances and exits.

If possible, counters should be low so as not to provide shoplifters with places to conceal their activities. As a general rule, store counters should run parallel, lengthwise to the location of the employee at the cash register. Then this employee can have an unobstructed view down the entire length of the aisle, providing another check on shoplifters.

How Shoplifting Is Done

It is difficult to generalize about the techniques of shoplifters. The methods they use to conceal or carry away merchandise are numerous.

An item may be simply picked up and worn out of a retail store, or carried as if the item were personal property of the shoplifter. This technique is used to steal gloves, scarves, sweaters, coats, or hats as well as leather purses or suitcases. A camera with a strap may be thrown over the shoplifter's shoulder.

In another situation, the shoplifter may go to a mirror, pretending to straighten out a hair arrangement, or a tie but actually dropping a small piece of merchandise down the collar or neck of whatever garment is worn. In still another situation, the thief may try several necklaces, unclasping one and allowing it to slip down inside the neckline.

The shoplifter may merely conceal a small item in the hand or cover it with a newspaper brought into the store. Others may drop items into the folds of a folded umbrella or slip merchandise into loose clothing. Ladies' handbags, baby strollers, and knitting bags are also frequently used as places of concealment.

Professional shoplifters may use coats with oversized pockets or "poaching coats" with hooks sewed inside a special lining. A great many items can be carried on these hooks without giving the coat's wearer an unusually bulky appearance.

Conditions and Factors Involved

There are a number of conditions and factors that play a part in shoplifting losses. More merchandise is now offered for sale in retail stores than at any previous time in history. Selling is often dependent on giving the customer an opportunity to handle the merchandise. Then, too, fashions and customs change constantly and many individuals feel that they cannot get along with the old items that they already possess.

Some authorities on shoplifting feel that it has increased because of a general decline in the morality and honesty of society. Some customers reason that the store's prices are already outlandish and that, in effect, the store owes the customer something. Students may engage in shoplifting as a sort of sport. Some housewives and elderly individuals shoplift to stretch their budgets. Almost invariably, these are persons who can afford to pay; few shoplifters are actually starving.

Good Customer Service as Major Factor in Reducing Shoplifting

A number of authorities on shoplifting agree that clerical attention to customers may be the most important single factor in controlling shoplifting. If clerical assistance is offered as soon as an individual enters the store, there may be little opportunity for shoplifting. Not only does an attentive clerk discourage shoplifters; such a clerk also pleases legitimate customers and can enhance the company's business. Most self-service discount stores trade off low clerical payrolls for high costs in the form of shoplifting losses.

Techniques for Displaying and Protecting Merchandise

If merchandise is arranged and displayed according to a regular pattern, an experienced sales clerk can tell at a glance if something is missing. Thus merchandise on display should be replaced from stock as soon as possible. In addition, glass display cases can protect some items against theft so long as the customer can see the merchandise being offered. Experience shows that other types of merchandise should be tied in place with cords or otherwise secured.

In a clothing store, it is generally advisable to keep clothes racks away from exits. Some shops button coats and jackets that are displayed on coat hangers, so that a professional thief cannot walk briskly into the store, grab twenty-five or thirty items, and run out the door. Some stores also alternate coat hangers, with the open hooks facing in opposite directions, so that a thief cannot grab a number of garments off the rack.

Surveys have shown that stores experience fewer thefts of small, "stealable" items when these are displayed very close to the cash register. In a supermarket, for example, items such as razor blades and flashlight batteries would be displayed at the check-out stand, while larger items would be displayed in other areas of the store.

When the Sales Clerk Is Not Certain

In most businesses, prosecution of all shoplifters is a desirable policy. The word gets around quickly and is an excellent deterrent. But at times, alternatives to prosecution should be considered.

If the shopper is a regular customer of the store and the sales clerk is not absolutely certain that something has been stolen, the clerk may use the so-called ghosting technique. If the sales clerk suspects the shopper of stealing a specific item, the clerk may obtain a duplicate item and place the duplicate in the shopping cart or among items to be checked out by the customer. No accusation is ever made, but the customer usually gets the point. Another possible technique is for the clerk to follow the customer closely; this will sometimes cause the customer to put the stolen item back on a shelf. This method also gets the message across while avoiding a confrontation.

Apprehension and Prosecution

Many security people feel that only a trained security officer, preferably a male, should apprehend shoplifters. Even though these offenders are usually harmless, they have been known to bite, scratch, or slug with their fists. Professional shoplifters sometimes have accomplices nearby who will come to their aid to brutally assault the arresting employee. In any instance where an arrest is to be made, the clerk who observes the theft should have another employee call any available security officers and members of management who may be in the store. Many establishments have the rule that the employee who observes the theft should continue to observe, the actual arrest being made by a security officer or ranking member of management, preferably a male.

It should be emphasized that when a person's guilt is in doubt, good judgment requires that the suspect go free. This is always the safest policy. The courts have repeatedly ruled that requiring the suspected shoplifter to return to the store constitutes an arrest, even when physical force has not been used.

Court judgments in a false arrest suit are usually very costly. Experienced security officers usually feel that a shoplifting arrest should never be made unless one clerical employee has been able to observe the suspect without any interruption whatsoever. If the suspect has not been kept in sight constantly, he or she may have passed the stolen item to another shoplifter and thus a search at the time of arrest will turn up no evidence.

Most prosecuting attorneys prefer that arrests be made outside the store or building where a theft takes place. Legally, a theft has taken place if a suspect leaves the premises with an item without paying for it. The fact that a suspect has gotten completely out of a store carrying unpaid-for goods will usually be of greater help in convincing a jury of the suspect's intent to steal the merchandise.

On the other hand, in the interests of protecting a firm against lawsuits for false arrest, experienced security officers advise that a suspect never be accused outside the store but be asked to return to the office to discuss a matter with the manager or the security officer.

If an arrest is made, a search should always be undertaken to recover the stolen items. If the suspect will not give up the property willingly, the items may be forcibly taken. A shoplifter should always be searched by a person of the same sex as the shoplifter.

The stolen items should be immediately marked at the time of recovery. The initials, place, and date should be written on each recovered item or on a tag affixed to each item by the employee or security officer who made the search. If an arrest is made, the security department must prosecute unless an admission is received from the suspect. Failure to prosecute may open the store to a subsequent lawsuit for false arrest.

The police should also be called immediately, to take charge of the arrested person.

After the arrest has been accomplished, the security officer should fill out cards pertaining to the facts of every arrest. These should be retained on file indefinitely. If the suspect should be arrested again at a later date, the thief will not be able to ask for leniency if a record has been prepared reflecting the first arrest.

It is also helpful for neighborhood stores or stores in a shopping center to maintain a central list of shoplifters. This list should be available to all cooperating merchants, but the maintenance of this list can prove to be costly if the list is not closely guarded. If this confidential information is not controlled, the individual whose name is on the list may be able to recover a judgment in a lawsuit if publicity is given to an arrest record.

Some Problems in Clothing Stores

Clothing items are unusually high on a shoplifter's list of desirable merchandise. Various methods have been designed to thwart potential theft.

Security coat hangers allow a customer to examine items such as leather jackets, expensive furs, and suede coats without removing the merchandise from the hanger. The hanger locks onto the garment, but the lock can be released by the sales clerk if the customer decides to try on the item.

Specially manufactured security garment hangers are also helpful in controlling removal of clothing from racks. Some hangers of this type are made with looped shafts so they cannot be removed from the rack. Others are manufactured so that they hook into a garment rail that locks into position on a rack. A number of other restraining devices and rack alarms are also available to control expensive merchandise.

Electronically activated price tags are also available to replace conventional price tags for merchandise. These electronically activated tags cause a matching sensor to send out an alarm if the tagged article is brought within a specified range of the sensor. Sensors are placed at exit points such as doorways, elevators, and stairwells. A person who goes through one of these points carrying merchandise with the tag still on it will set off the alarm. When an item is paid for, the tag is deactivated and removed by the clerk at the cashier's stand. A system of this kind is very effective but maintenance is expensive. In addition, there is always the possibility of embarrassment to a customer if the merchandise tag should not be deactivated by the clerk at the time of purchase. If this should be the case and an arrest should be made on the supposition that the item has been shoplifted, a false arrest suit could result.

Shoplifters frequently carry items of clothing into fitting rooms in clothing stores and then come out wearing the item as if they owned them. To control thefts from fitting rooms, some stores allow only one customer or one person in the fitting room at one time. Others control the number of items of clothing that can be taken into the fitting room by an individual customer. Clerical employees oversee the system to be sure that items taken into the fitting room but not purchased are returned to the sales floor. Some stores give the customer a coat hanger with a colored plastic disk on which to hang the clothing items. The color of the desk designates the number of items for which the customer is responsible.

When shoplifters wear stolen items from a clothing store, they almost invariably remove price tags or tickets. These tickets are frequently secreted behind seats in the fitting room or placed in a waste basket. A quick check of the fitting room for discarded garment tags or tickets may sometimes alert the sales clerk to a theft.

If clothing items were given to the customer in a box, the sales clerk should immediately verify that the contents have been returned if the items are not purchased.

Better control of fitting rooms can be accomplished if they are kept locked when not in use.

Some Other Protective Techniques

Retail stores utilize alarm systems in some unusual applications to protect merchandise on display or in areas where it may be available to shoplifters. For example, an antitheft cable alarm system can be strung through the handles of a number of pieces of merchandise that are located in the same general area. This alarm consists of a lone cable, quite long, that will activate a response if the cable is cut. A system of this kind uses an electrical current that travels the entire length of the wire cable through handles or special fasteners on all of the merchandise under protection. The electrical current returns to the unit, completing a circuit, and any tampering that breaks the electrical circuit will set off the alarm.

Another type of alarm system is designed with a strip of separate junctions fastened along the back of the counter and connected to a uniform alarm source. Yet another type operates on the basis of negative pressure sensitivity; if items stored on display stands or shelves are picked up by a thief, the alarm will sound.

A variety of other preventive measures have been used to detect shoplifters. Store detectives sometimes conceal themselves and use lookout points or two-way mirrors to observe customers' activity. Some stores rely on concave mirrors to give the sales clerk a better view of activities in distant parts of the store. Mirrors can be of definite value in observing shoplifters, but many mirrors are no more than psychological deterrents.

CCTV cameras are another tool for crime prevention, and some retail stores mix "dummy" cameras with live equipment, concentrating attention on those parts of the store most susceptible to shoplifters.

Frequently stores post signs to remind customers that they will be prosecuted if they are caught shoplifting. In the past, some stores were reluctant to offend customers by implying they could be dishonest and stopped using signs. But because of mounting losses in recent years, more and more stores have posted warning notices.

Department stores sometimes use women security officers to actually observe women customers dressing in fitting booths. The disadvantage of this policy is that many customers, knowing they were being observed, would be offended. In most instances, however, the courts have held that evidence collected through observation by a woman security officer under these circumstances is admissible in a prosecution for shoplifting (theft).[1]

---

[1] See People v. Victoria Randazzo, 220 Cal. App. 2d 768.

Educational Programs in High Schools

Shoplifting by high school students can sometimes be sharply curtailed by informative lectures to students by local police officers or representatives of the Retail Merchants Association. In lectures of this kind, it is helpful to point out to students that they may be handicapped in obtaining jobs in the future if arrested for shoplifting, that their future security clearance for governmental jobs could be jeopardized, and that bonding applications or college admissions could also be affected. Appealing to the students' sense of fair play is also helpful in discouraging shoplifting among high school groups.

SUMMARY

Shoplifting is not restricted to any specific class, age group, or race, although juveniles and housewives bear considerable responsibility.

To reduce losses, it may be helpful to design store interiors with a minimum of blind spots or large pillars. Designs without L-shaped sales floors are preferable, and there should not be so many exits that they cannot be controlled. A balcony that offers a view of the sales floor below is a good deterrent to shoplifters.

Experience indicates that checkout stands and cash registers should be located near exits. Shoplifters cannot hide behind display boards if they are located high on the walls of the sales floor. Acrylic, see-through fixtures also improve employees' ability to observe suspected persons. In so far as practical, merchandise counters should be low, running in parallel rows that permit a sales clerk to look down the aisles from the checkout stand. Small items of considerable value should be located close to the checkout stand.

Unless protective measures are taken, items such as suitcases, purses, scarves, and sweaters may be simply picked up and worn out or carried from the store.

Some firms have found it helpful to keep counter display compartments full, so that the absence of an item will be immediately spotted by an alert sales clerk.

But regardless of the preventive steps that may be taken, authorities on shoplifting are in general agreement that prompt clerical attention to customers may be the single most important factor in combating this kind of theft. If a large number of high school students pour into a store immediately after school lets out, scheduling extra clerks for this period may reduce shoplifting.

Clothing stores may have more than their share of problems related to shoplifting. Some stores utilize special hangers that prevent easy removal of garments. Reversing the hanger on every other garment on a rack is another preventive technique, making it difficult for a thief to grab a number of garments and bolt out the door.

Varied types of alarms are available to protect valuable items. A TV set, for example, can be plugged into an electrical outlet with an alarm that sounds loudly if the plug is removed. Electronically activated tags can be placed on garments, setting off an alarm if the garment is carried past the checkout stand before the tag is inactivated by the employee at the cash register.

Garment fitting rooms can be locked between customers and a limit set on the number of items given to a single customer. Inspection of fitting rooms as a customer departs may uncover shoplifting if discarded price tags are found in that area.

Many merchants and police authorities feel that prosecution of all shoplifters is an effective deterrent. But if a manager or security officer arrests a person who is not guilty of shoplifting, a very costly false arrest lawsuit may result. Employees should therefore be certain that a theft has actually been committed before an arrest is considered. It is important for the witnessing sales clerk to keep the thief under constant observation until the arrest is actually made. If observation is interrupted, even briefly, the thief may pass the stolen item to a confederate so that no stolen property will be found at the time of arrest. Experience also indicates that a shoplifter should seldom be arrested by a lone woman employee, without someone from security or management to assist in the event physical help is needed.

Legally, a theft has been committed when merchandise is taken with intent to steal, but most prosecutors feel that the case against a thief is made much stronger if the thief is apprehended after leaving the building.

Experienced security people generally agree that the suspect should not be accused outside the store. If the suspect will not return to the store voluntarily to discuss the matter, the arrest must be made outside and the suspect taken back to the office. When an arrest is in fact made, the suspect should be immediately searched and the stolen property recovered and held in file for the prosecutor. Each item of recovered merchandise should be initialed, dated, and marked as taken from the suspect. The suspect should then be interviewed and asked to sign an admission of theft. The police should be called to take the arrested individual into custody, if prosecution is to follow. A signed admission of theft or a conviction in court is the only real protection against a lawsuit for false arrest.

Educational programs in local high schools, handled by local police officers, are sometimes very effective in reducing shoplifting incidents.

REVIEW AND DISCUSSION

1. Is shoplifting usually committed by one specific age group or economic class? Explain.
2. What can be done in the design of a building to reduce shoplifting?
3. Describe some of the techniques commonly used by shoplifters.
4. What is the single most important business activity or factor that may be used to control shoplifting?
5. In laying out a store or merchandise display, what are some techniques that may help clerical employees in controlling shoplifting activities on the sales floor? Describe the ideas you are talking about.
6. Outline training that may be given to store employees to avoid giving shoplifters grounds for false arrest lawsuits.
7. Should the shoplifter be arrested inside the store? Why?
8. Should the shoplifter be searched when arrested? What should be done with stolen merchandise that may be found on the shoplifter?
9. Describe some protective features or techniques that may be used by retail stores to protect their merchandise.
10. Describe how merchandise items should be marked to prevent shoplifting or ticket switching. How are losses best controlled when the customer attempts to switch tickets?
11. If a signed confession is not furnished by a shoplifter upon apprehension, should a criminal complaint always be filed? Why?
12. What can be done to reduce shoplifting by counseling high school students?
13. Should the merchant always follow a policy of prosecuting shoplifters without exception? What are the benefits and possible objections to a policy of this kind?

9

Armed Robbery

The purpose of this chapter is to outline serious security problems that may result from business robbery. This material discusses how merchants, security officers, and police may cope with robbery in three basic ways:

1. Protecting employee safety.
2. Increasing the possibilities of identifying and apprehending those responsible.
3. Minimizing and recovering losses.

INTRODUCTION

Basic Differences Between Robbery and Burglary

In understanding robbery, it may be advisable to distinguish between robbery and burglary, since these two crimes are often confused in public thinking. Robbery is *always* a crime against a person. Frequently, it is a frightening experience for the victim. In the typical robbery, either one of two things happens:

1. The victim is overpowered and property is taken away by brute force. (This is a mugging or strong-arm robbery.)

2. The criminal threatens to use a dangerous weapon or do great bodily harm, and the victim turns over valuables because death or serious injury appears to be the alternative.[1]

In essence, burglary is an unlawful breaking and entering into a house, office, or building, with the intention of committing criminal acts after getting inside. A tramp breaking in solely to get out of the freezing cold would be guilty of trespassing but not burglary. In most instances the burglar wants to break in to steal. The burglar usually hopes to avoid being seen so there will be no confrontation with the victim.

There are, of course, some burglaries in which the criminal is under the influence of drugs and may not hesitate to attack anyone encountered after the unlawful entry. A burglary of this kind goes beyond mere breaking and entering to steal, involving other serious crimes. But when people say they "have been robbed," in many instances they have been victimized by a thief who burglarized the building.

At the outset, then, it should be pointed out that robbery usually involves far more potential danger for the victim than does the usual burglary.

Muggings and robberies of individuals have increased alarmingly in recent years. Most robberies, however, are directed at businesses, since individuals are less likely to be in possession of large amounts of money. But as an individual, a merchant may be unusually vulnerable to robbery at the time of departure from the store. This is because the merchant may be believed to be carrying large sums when leaving the business.

There are no absolute rules for avoiding personal robbery. It is wise to avoid walking alone, especially in poorly lit streets or alleys when there is little traffic or likelihood of observation from nearby homes or businesses. It may also be advisable to carry a loud whistle on a neck chain, or other protective devices.

---

[1]"Robbery is the felonious taking of personal property in the possession of another, from his person or immediate presence and against his will, accompanied by means of force or fear." *Black's Law Dictionary*, 4th ed. (St. Paul, Minn.: West Publishing Co., 1968), p. 1492.

"In essence, robbery is a theft by use of force. A common law felony from the earliest times in England, the offense is a felony in all American jurisdictions. The old English crime called "Highway Robbery" was simply any robbery on or near a public highway. Highway robbery has now acquired a broader meaning, being absorbed into robbery in general. . . . The object taken must be personal property, capable of being stolen . . . taking must be with the intent to steal . . . must be from the person or from the immediate presence of another person . . . it must be taken . . . either by violence or by threats. . . ." A.Z. Gammage and C.F. Hemphill, Basic Criminal Law (New York: McGraw-Hill, 1974), pp. 237–242.

Business Robbery

The likelihood of business robbery is increased whenever valuables (cash, merchandise, stocks and bonds, etc.) are stored or transferred.

It should be pointed out that most business robberies involve the use of a gun, knife, bomb, or other dangerous weapon. For a number of years, armed robbery has ranked as one of the five major causes of lost revenue for supermarkets, grocery stores, and small shops. But many other types of businesses may also be victimized. Many companies, therefore, feel an obligation to prepare those employees who may be confronted with this situation. At the least, cashiers, tellers, and other employees handling money should understand what may happen.

In coping with robbery, three basic objectives should be kept in mind:

1. Maintain maximum employee safety.
2. Increase the possibility of identifying and apprehending the criminal.
3. Recover losses that may be involved.

Authorities on business robbery are in general agreement that apprehension and certainty of punishment will deter an appreciable number of armed robberies.

Armed Robbery Is Not Always a Logical Crime

Professional criminals will usually participate in a holdup only when the anticipated loot and the chances for success appear to overshadow the risks involved. Generally, the real professional weighs a number of factors—whether there is likely to be determined resistance, whether the police will arrive on the scene in a short time, whether a great deal of money can be obtained, whether neighbors are in a position to interfere with the crime, and whether the bandit may be forced to shoot someone to make a getaway.

The inexperienced holdup man, however, is not always motivated by logic. This kind of bandit may rob whenever short of funds, sometimes seeming to act on mere impulse. As a class, dope addicts may also completely ignore logic, pulling a robbery to satisfy the dictates of an expensive drug habit.

It is therefore apparent that businesses cannot always avoid robbery, even though reasonable precautions are taken. There is much that can be done, however, to protect employees and to minimize money losses.

Almost invariably, those businesses that remain open after normal closing hours are more vulnerable to robbery. Bars, service stations, neighborhood markets, motels, and liquor stores all seem to have this extra risk.

From time to time, articles appear in the news to the effect that an angered shop owner or store clerk overpowered a holdup man with a tire iron, an item of merchandise, or whatever weapon of defense was found handy. These news accounts make for interesting reading but often reflect questionable judgment. Most experienced peace officers believe that resistance should be offered only under extreme circumstances.

Usually, the orders of the robber should be complied with as closely as possible. Any movements made should be careful and deliberate. Nothing should be done to excite the robber or to create the impression that the victim will resist. Although the victim may be able to avoid personal harm, bystanders may not be so fortunate. Employees should be taught that it is better to be a live witness than a dead hero!

Instead of resisting, it may be more important for the victim to concentrate on remembering accurately everything he can about the robber's appearance, dress, method of operation, voice, and mannerisms; if a getaway automobile is used, the victim should try to observe as many details about it as possible.

BEFORE THE ROBBERY

It is not always necessary to spend large amounts of money or time in planning to avoid robbery. What is needed, however, is an alert management and the development of security awareness among employees.

Many firms have found it helpful to give brief training periods to key employees. In a session of this kind, employees can develop their skills in observing and recording significant details. They can learn to judge the height of a criminal by noting where specific features correspond to objects on the wall or the cashier's cage. The security officer can explain what to expect and how employees should react.

Periodic security reorientation sessions with employees should also be considered to make certain that there is a continuing awareness. If handled in a serious but low-key fashion, these sessions can be conducted without causing undue alarm among employees.

It is suggested that robbery procedure cards or "tickler" cards be given to supervisors and managers at training sessions. If most of the firm's money is retained in an on-site money room, these cards should also be given to cashiers, tellers, and money room employees. Instructions on these cards will, of course, vary from location to location, depending on a number of factors.

A typical robbery procedure instruction card would include instructions like the following:

1. Take no action that will harm yourself or others!
2. Set off the company alarm.
3. Put the surveillance camera in motion.
4. Place "bait money" in the loot.
5. Observe the bandit's description, dress, and equipment.
6. Call the police, whether or not the alarm was set off: Telephone # 643-8100.
7. Protect any evidence left, touched, or dropped.
8. Observe how the getaway was made; record the license number.
9. Jot down names of customers and bystanders who may be witnesses, along with their addresses.
10. Pass out blank paper and ask witnesses to write down everything observed, including descriptions.
11. Station someone at the door to let arriving police know the bandit has left.
12. Keep curious and unauthorized persons out until police arrive.
13. Notify company auditor.
14. Answer press inquiries.

A robbery instruction card of this kind can be kept in the employee's billfold or wallet or in a desk drawer where it is not likely to be taken along with holdup loot. Some prefer to keep this information in credit-card size, prepared on a copy machine with reduction capacity.

Recognizing That the Business Is Being "Cased"

Although crimes are often committed on impulse, experience shows that comparatively few robberies occur without at least some prior examination of the premises. At times, the bandit will merely come in a store and look at merchandise without buying anything, or perhaps purchase a small article. In a robbery involving several persons, it is not unusual for more than one of these to insist on "casing" the building.

If a customer's dress and grooming seem out of place in a particular business, this may be a tipoff that the individual is a potential robber. From time to time, any store or shop will have visitors who simply do not fit the pattern of most legitimate customers. Some of these persons may be buying an item about which they appear to know nothing, or some may seem ill at ease. Employees need not surmise that all nonconformists are criminals, but any unusual situation should at least put employees on the alert.

Some bandits may be able to fit in to the store's atmosphere so naturally that they attract no attention. And sometimes a robbery may be set up through inside information from an employee or regular customer, so that the bandit does not feel compelled to come in beforehand. This is usually the exception.

If clerical employees observe these indicators, or sense that the business is being set up for a robbery, an immediate call should be made to the police department or sheriff's office covering the location. Police agencies will usually dispatch officers to the neighborhood, if they are available. A stakeout in a situation of this kind will often result in apprehension of individuals who have just committed a robbery.

Recently, in a California jewelry store, a sales clerk concluded that the firm was being "cased" for a robbery. As soon as two suspects left the store, the clerk alerted a second employee. While the first clerk called the police dispatcher, the second walked casually out the front door of the building and noted the license plate of an automobile parked down the street, with the engine running and a driver at the wheel.

Needless to say, the suspects returned to the store within a matter of minutes, perpetrating an armed robbery. Provided with the license number of the getaway car by the alert clerk, plainclothes police officers arrested three participants as they fled from the scene.

Obviously, employees should not become overly involved in playing detective. But alert employees, who have been briefed in advance, can frequently sense a situation of this kind before it develops.

Some firms have installed dummy TV cameras, hoping to deter criminals while saving the cost of real cameras. This type of installation is of doubtful value, since employees may feel that management is not willing to provide them with the protection they should really have.

Some Precautions Beforehand

There are other precautions that will usually reduce the likelihood of robbery. For example, it is preferable for at least two employees to be present when a business is closed for the day. Experience shows that a lone employee may be "jumped" just after other employees have departed.

Cashiers or checkers should not be allowed to balance their own cash in an area that can be readily observed by outsiders. And some thefts can be eliminated by being certain the cash register drawer is invariably closed after each transaction.

If the cash register is relatively light in weight, it should be bolted to a table or counter. Occasionally, a strong bandit grabs up a register, carrying it to a waiting car before employees can recover from the surprise.

Armed, uniformed guards who are trained and supervised are very effective in providing actual protection, as well as serving as a psychological deterrent to robbery.

Kidnapping, Coupled with Robbery

Kidnapping of a business executive or store manager is always a distinct possibility if the kidnapped victim is in possession of the combination to the safe. In a recent Michigan case, two bandits posing as policemen made off with approximately $90,000. They accomplished this by holding a store manager's family hostage while he was forced to remove weekend receipts from the company strongbox.

The bandits gained entry to the victim's home by claiming to be police officers. One of the criminals then remained with the manager's family at the home, while the second forced the manager to accompany him to the store and work the combination to the safe. Messages were then exchanged between the bandits by walkie-talkie radio. The manager was returned to his home, and he and all members of his family were handcuffed and left in a bedroom closet.

A situation of this kind can usually be avoided if the victim and members of his family are very careful to determine the identity of anyone who seeks admittance to their home. Requesting emergency use of a victim's telephone is a favorite way by which bandits gain entry.

Some business safes especially designed to avert this type of robbery are equipped with a dialing combination that permits the safe to be opened, but simultaneously sends a secret alarm to a police station or central alarm facility. Other safes make use of a time clock arrangement so they cannot be opened during the night.

A Reward Program Is Usually Ineffective

Posting a reward in advance has usually been ineffective in preventing business robberies. A reward offer posted after a robbery may sometimes bring results; however, it is usually effective only if the reward is a considerable sum of money.

MINIMIZING THE LOOT

It seems obvious that a bandit should never be given more money than he asks for. Yet cashiers sometimes hand over all the money in the drawer when only a moderate sum has been demanded. One way to control this type of loss is to counsel employees before a robbery takes place. As a general rule, the most effective way to minimize the loss in a business robbery is to hold down the cash accumulation rather than to hold out on the bandit.

Two approaches to avoid accumulating cash may be taken. One is to deposit incoming receipts of cash in the bank on a daily basis at least. An armored car service is better to use than a company courier. Deposits may be required more frequently than once a day if there is a substantial buildup of money during the Christmas shopping season or when business exceeds expectations.

A second approach is to make certain that employees still on the premises do not have access to funds after hours when they may be confronted with a holdup man.

If large sums of money are needed for change or for cashing customers' payroll checks, then armed guards or other stricter than usual security measures may be warranted.

Setting a Cash Limit

Criminals sometimes make errors in casing a robbery, as illustrated by a recent Los Angeles holdup in which the criminals obtained over $1 million in checks but no cash or negotiables. Usually, a professional robber will not attempt a crime of this kind without reasonably good information that a substantial amount of money will be available.

Robbery will be discouraged by letting the criminal element know that a money limit is closely adhered to. The opposite result can be expected from employee activities that "advertise" the retention of cash on the premises. Frequently, police officers report that bandits were encouraged to perpetrate a supermarket holdup or bank robbery only because employees carelessly displayed large sums.

Bleeding Cash Registers

It is essential for managers of retail stores to bleed cash registers frequently. In addition, employees should be taught to avoid discussions of cash receipts or money transactions in lunch rooms or public places. Armored car pickup schedules or bank delivery routes should not be disclosed.

The amount of cash accumulated will, of course, vary considerably from company to company and from store to store. Increases can be expected on certain days of the week, around holidays, and after payday in industrial areas. Management should study these factors and set up definite rules governing the amount of money that may be carried over from one business day to the next.

An open display of a cloth money bag enroute to the bank may encourage an armed robbery or a "grab and run" attempt. It is usually advisable to carry the bank bag in a briefcase or plain paper sack. Transporting money by car rather than on foot should also be considered, especially if there is an armed guard to cover arrival at the bank's parking lot.

As noted previously, use of an armored car service may be the ideal solution to prevent money accumulation in a business office. But it may be a serious mistake to turn funds over to any individual wearing an armored car service uniform. Anyone wearing the uniform who is not recognized on sight should be challenged for proof of identity. There are a number of cases on record of impostors who wrongfully obtained large sums of money by wearing an old armored car messenger's uniform.

Protection Should Also Be Given to Checks

Most thieves, of course, are looking for currency rather than business checks. Occasionally, an unprofessional criminal may try to cash a check that he has obtained in the loot from a crime. For this reason it is advisable to place stops with the bank against the payment of any checks known to have been taken in a robbery or burglary.

It is a good business practice for the cashier to mark all incoming checks "For Deposit Only" at the time they are brought into the company office or money room. If checks that bear this stamp should be subsequently lost or stolen, any bank cashing the check will then be legally responsible for the loss.

But the real problem here is that checks are often included in the loot, prior to being stamped, or prior to a record of each check by maker. The criminal simply scoops up whatever loot is available, without separating checks from greenbacks.

In the typical case, the checks taken in a robbery are eventually burned or thrown away. But in many instances the firm that loses the checks has no way to determine who gave the checks and who should be contacted to replace the ones taken by the holdup man.

Location of the Company Money Room

There is a difference of opinion among business officials as to where the firm's money room or cashier's cage should be located. Some feel that this room should be placed within the public view, so that any unusual activity would be readily observed by a number of employees or by the general public.

Most businessmen and security professionals, however, believe that the money room should be in an isolated, inner area, where there is little traffic involving nonemployees. In a location of this kind, an outer area with a locked door will give additional protection to the money room. This is especially true if the outer door remains locked and equipped with a peep hole, and if there is an alarm system that can be activated in an emergency.

Bullet-resistant glass windows in cashier's cages have been found quite effective. Such barriers should be at least 1 3/16 inches thick, made of a good grade of resistant glass, and have walls of steel plate. Pass-through devices in the cage should be constructed in such a manner that a criminal outside the cage cannot obtain a direct line of fire at the occupant of the cage.

The Holdup at Opening Time

Almost invariably the criminal will have better control of activities if able to rob a business when it is still closed to the public, with only one or two employees present.

If a store or business retains considerable cash on the premises overnight, there is a possibility that a holdup man may get inside during the night by breaking and entering. The bandit may then stay hidden and surprise employees as they come to work. This kind of robbery may be very difficult to prevent if the first employee has the combination of the safe and there is an appreciable time lag before a second employee reports for work. If the first employee coming into the building does not know the combination, the bandit may force this employee into an office closet, awaiting the arrival of one of the officials who does have the combination. In some instances, a holdup man may force ten to fifteen employees into a closet, or tie them up one by one, while waiting for the employee with the combination to appear.

In a variation of this kind of robbery, the criminal may not use a burglary technique to gain entry into the building. Instead, movements may be timed to pull a gun on the first employee, just as the employee places a key in the door. Among the criminal element, this kind of opening-time robbery is called a morning glory holdup.

This type of crime can usually be avoided if employees do not enter the store in a group. Some should remain outside, in case a robber tries to "take" someone as the employee enters the business premises.

The first employee to arrive should wait to enter until a second employee arrives. The second person can then wait outside while the first checks to see no one is concealed in the building. If it is safe to enter, this person can give a prearranged signal to the employee outside. If the "all clear" signal is not given, the employee out on the street should immediately call the police.

Paying by Check

There are few industries in the United States where employees are not paid by check. If employees insist on cash, it is suggested that consideration be given to staggering paydays, so that the cash accumulation is not a great incentive to robbery.

WHAT MAY BE DONE DURING THE ROBBERY

Business employees who have given advance thought to robbery frequently report that planning helped them to remain cool during an actual holdup. It serves no purpose to panic, and it may seriously alarm the bandit if the victim screams.

Some authorities on robbery feel that the victim should immediately indicate to the holdup man he or she intends to follow instructions. In some cases, the criminal may be even more nervous than the victim. Assurance that there will be compliance with the robber's demands may decrease the danger to the individual being robbed, as well as to other persons in the vicinity.

Tripping the Silent Alarm

If there is a silent robbery alarm system and/or a camera surveillance system, it is recommended that the controls be activated just as soon as it is apparent that a robbery is taking place. This, of course, cannot be done if the controls are located where this action may be observed by the stickup man.

Some businesses follow a policy of not activating the alarm until the bandit walks out the front door. Most authorities on robbery feel that there should be no delay, however.

Exhaustive statistics on robbery reports where the victim was harmed are not available on a nation-wide basis. It can be said, however, that in the great majority of instances, there is no harm to the victim who does not resist. There are isolated cases, however, in which the bandit may seek to hurt the person robbed, even though this individual has done nothing to motivate or trigger the attack. In instances of this kind, obviously, the sooner the call for police assistance, the better.

Employees of some stores or shops, fearing the loss of their own money, and knowing that they have already sent a silent alarm signal, sometimes decide to grapple with the holdup man. They assume that help will arrive soon and that they can contend with the criminal for a few minutes. What they fail to consider, of course, is that the police department may have already dispatched all available units on other emergency calls, or that a police unit may be unable to get through traffic in time, even with a red light and a siren.

At other times, the victim may conclude that the gun being displayed is merely a toy pistol and the bandit can be subdued without much risk. But sometimes a real gun is mistaken for a toy and the consequences of this mistake may be tragic.

Call Police Authorities Not the Company Auditor

Time is of the essence in solving armed robbers' violations, and it is essential that the police be notifed as soon as possible. Just because a silent alarm has been activated, it should not be assumed that the report has reached the police dispatcher. Police should be contacted by telephone in all cases. To facilitate this call, it is advisable to have the police emergency number posted where it is easily seen. Some businesses place this number on a bulletin board, along with emergency numbers for the fire department and ambulance service.

Experience also shows that far too frequently employees will make a call to the wrong police agency. For example, the business that is robbed may be located outside city limits, where jurisdiction for criminal violations rests with the county sheriff rather than with the city police. Considerable time may then be lost while the agency called relays the information to the proper law enforcement department.

And while it seems logical to advise police immediately, experience also shows that persons having responsibility for company money will sometimes telephone the company auditor's office first, advising that official of the robbery. Notifying the proper police authority first should be stressed to all employees.

Another mistake business employees frequently make at this stage of the crime is to get the police agency dispatcher on the phone, hurriedly report the business's name and address, and then hang up immediately.

If possible, the line to the police dispatcher should be kept open and a company employee should remain on the line until told to discontinue. This enables approaching police cars to be immediately advised of new information as it is received. Any details furnished by neighboring businesses or witnesses who were passing by can then be relayed—which way the bandit fled, the description of the getaway car, etc.

The employee holding the open line to the police agency should understand the increased likelihood of a quick apprehension if patrol units can approach with good descriptive information. Then, too, there may be a limited number of telephone lines into the business, and if they all become busy, the police dispatcher will be unable to call back for more details.

Giving a Signal to Co-Workers That a Robbery Is in Progress

In some businesses it is possible to give a signal to a co-worker that a robbery is in progress. In a market, for example, a bandit may point a concealed gun at a checker in such a way that it cannot be seen by other employees in the store, even though they are close by. Through a prearranged signal, the employee being robbed may be able to alert a co-worker as to what is taking place. The second employee, if unobserved, may be able to slip out and call the police or give a signal to still another employee.

A system of this kind requires alert employees, along with a measure of teamwork. When the system functions properly, the bandit may be taken into police custody without even realizing that employee action caused his apprehension.

As a matter of policy, it should never be made known to the holdup man, even after arrest, that the police were notified by someone in the business that was robbed.

The Need for an Accurate and Complete Description of Each Bandit

The police can often be successful in apprehending a holdup man only if they have a good mental picture of the person they are looking for. Obviously, this mental image must come from the victim or nearby witnesses who observed what happened.

The ability to describe a criminal accurately can be developed. A witness's ability can usually be increased considerably if the company security officer or someone in management will spend a few minutes going through a mock robbery. A cashier or teller can learn to estimate the height of a person standing in front of the money drawer by judging from preselected marks drawn on the wall. After some trial and error, the person can develop the ability to do this without the use of height markers.

If the witness will remain calm and collected, a wealth of descriptive detail can be gathered in a few seconds. This is best done by making a methodical, orderly inspection of the bandit's facial features, coloring, accent, race, age, sex, scars and marks, coloring, wearing apparel, speech, walk, and any other features.

As in judging height, brief practice sessions will enable employees to note a number of significant details that may help the police to identify the culprit.

Some details of an incident may remain clear in a person's memory throughout an entire lifetime. Other impressions may fade within a short period. It is therefore important for witnesses to record descriptive information as soon as they can conveniently do so. A form for this purpose appears in Appendix 3.

Witnesses often get together and compare impressions after a holdup. Experience shows, however, that some witnesses seem to dominate others. Frequently, a witness will substitute information overheard from another witness instead of reporting what he himself actually observed.

The independent impressions and opinions of each witness are what must be obtained. And the sooner this information is recorded, the more accurate it is likely to be. If one witness colors the impressions of another, the police may form an erroneous picture of the person they are looking for.

As soon as the robber leaves and the police have been notified, witnesses should be furnished a description form or sheet of paper and required to fill it in independently. The office copy machine can be used to quickly run off copies for each witness that may be involved.

Unusual Features of Bandit May Be Significant!

It is the unusual features, or peculiarities, of the bandit that should be given most attention. A noticeable scar, tattoo, mole, blemish, or birthmark may enable the police to solve the case quickly. Any peculiarity may also be very useful in developing suspects if the solution is not reached in a short time.

If the bandit's appearance is not unusual, then there may be significant peculiarities in dress, jewelry, eyeglasses, type of watch worn, or other personal items.

In a recent case in the Southwest, a witness related to police that the man who demanded her money at gunpoint had a plastic pencil holder in his left shirt pocket. The witness was also able to recall that there were three drafting pencils in his holder, one red, one green, and one purple. Alerted to this information, a police patrolman arrested the culprit when the officer observed a man with this combination of colored pencils in his shirt pocket. When apprehended, the criminal was purchasing a bus ticket to get out of town.

Observing the Bandit's Activities and the Objects He Touches

Persons who observe a robbery should be encouraged to note everything that the bandit touches, as well as to report each action, before and after the crime. If the bandit comes into contact with fixtures, doors, windows, paper supplies, or merchandise, it may be possible to lift fingerprints from some of these items. Witnesses should also take steps to prevent other people from touching or trampling on these items until the police officer in charge of the investigation has had an opportunity to decide whether evidence could be obtained from the items in question.

A holdup man may frequently discard some item of significance in fleeing from the scene of the crime. A cap or hat that falls from his head, for example, may contain human hairs along the inner liner that are of value for comparison by the police laboratory. Or discarded gloves may be found by the police laboratory to contain substances or chemicals that give a clue as to the type of industry or trade in which they were used prior to the robbery.

It is not unusual for a holdup man to wait outside a cashier's window for an opportune time, pretending to write a check or credit application form. This may be discarded in the wastebasket shortly before the bandit pulls out a gun. A paper specimen of this kind could be valuable to the police laboratory for handwriting comparison, for fingerprints, and for ink comparison.

The victims of a robbery are not expected to serve as investigators or detectives. But it should be emphasized that they render a considerable service if they observe and protect evidence until police arrive. Although it is difficult to predict just what may turn out to be evidence, the question should be resolved by presuming anything the bandit used, touched, or discarded could fall into this category.

To illustrate what may be done with evidence, consider the case of a recent holdup of a cashier in a department store. The bandit passed a threatening note through the cashier's window, pulling back his coat long enough for the cashier to observe that he carried a gun tucked inside his waistband.

The cashier complied with the robber's demand for money and had the presence of mind to retain the criminal's note. A suspect was developed by the police a short time later.

1. Tear marks at one end of the paper note were examined microscopically. The tear marks made a perfect match with the remnant of a sheet in a paper tablet found in the suspect's room.
2. The handwriting on the bandit's note was identified by a handwriting expert as that of the suspect, according to the examiner's opinion.
3. Fingerprints on the note were identified as those of the suspect.
4. Spectrographic analysis of the ink samples on the note reflected the same chemical composition as fluid in a writing pen found in the suspect's shirt pocket.
5. Analysis reflected the paper on which the note was written had the same weight and watermark as the paper in the suspect's tablet.
6. When the bandit wrote the threatening note in the tablet, the pressure made indentation marks on the sheet of paper immediately below. These indentation marks were brought out by laboratory techniques. The indentions in the suspect's tablet corresponded exactly with the writing on the bandit's note.

It is highly unusual, of course, for one small piece of evidence to prove so damaging to the criminal as the note in this particular case. But witnesses should make every effort to protect any article that may possibly lead to the identification of the criminal.

Giving Out the Bait Money

A case can almost always be proved to the satisfaction of a trial jury if the money taken at gunpoint can be identified by serial number. In police language, currency with recorded serial numbers is known as bait money, or marked money.

If bait money is kept on hand, the employee who is held up should make certain that this money is included in the loot taken. As a practical matter, it is usually troublesome to maintain bait money in a cashier's cage, as it may be given away unintentionally in making change. Bait money must be kept in such a condition that it is not obvious to the bandit. If the money is retained separately, and marked in any noticeable way, the holdup man will probably flush it down the first available toilet.

Ideally, the cashier preparing a bait money list should record serial numbers of bills in handwriting and should date and initial the list. The cashier should not retain the list in a cash drawer, as it may then be taken with the loot. Generally, it is preferable for a bank bill strap (money wrapping band) to be placed around the bait money, so the cashier can distinguish it from other currency. But some other money should also be retained with bill straps, so that the marked currency is not distinctive.

There is no absolute legal requirement that the bait money list must be prepared in the cashier's own handwriting, but this provides a valuable piece of evidence in a criminal prosecution against a holdup man. Considerable experience in criminal prosecutions indicates that this is the best method to use.

It should also be pointed out that the serial numbers on U.S. currency appear on only one bill of the series in which issued. This same serial number, however, may be repeated when bills are issued in a new series year. Therefore, the bait money list should include all the letters and eight digits of the serial number, along with the series year of issue. If an asterisk (*) appears in place of one of the alphabetic letters in the serial number, this asterisk should also be recorded on the bait money list.

AFTER THE ROBBERY

It is of considerable importance for the robbery victim to observe the getaway route, as well as details about the getaway vehicle. In many cases a robber ducks into a neighborhood bar or nearby business, expecting that this will not be observed. At other times, the criminal melts into pedestrian traffic on a busy street, believing this is the best way to shake off pursuit. And every year there are a number of cases in which victims or onlookers have managed to follow just closely enough to be able to point out the guilty man to the approaching police.

In cases where a getaway car has been used, witnesses often were able to get close enough to read all or part of the license on the vehicle, or at least provide a description of the car model and its color.

A stolen vehicle or stolen license plates are frequently used in major holdup cases or in bank robberies. But in a good percentage of all armed robberies, the bandit uses his or her own vehicle. At times a bandit may try to disguise the vehicle with a stolen license plate or by covering the license with mud, but in most instances a criminal relies on parking some distance away to prevent identification.

On occasion, a witness may get close enough to the bandit's car to read the license number but has no writing materials with which to record it. In one recent case, the witness scratched the license plate digits in some roadside mud with the branch of a tree. In another incident, a woman wrote the vital numbers on the sidewalk, using her lipstick as a crayon.

Recording Names and Addresses of Witnesses

The impressions of all witnesses may be vital. In a retail store robbery, a customer who witnesses the crime may not recognize the need to stay on the scene until police arrive. There is no way that a store manager can force witnesses to remain, but someone in the store should be designated to record names, addresses, and telephone numbers of possible witnesses. This list should be furnished to police, so that necessary contacts with these individuals can be made.

If a customer-witness declines to furnish name and address, then if possible the license number of this person's vehicle should be recorded on the list. The police can identify the individual from this information.

If any cash was overlooked or dropped by the fleeing bandit, a supervisor should immediately take steps to make certain that this money is safeguarded. Unsupervised victims of a robbery have been known to steal funds that were left, causing a greater business loss than the actual holdup.

Advising the Police That the Robber Has Departed

One of the most critical moments in any armed robbery is the time when the first policeman arrives at the business door. This is when a cornered bandit may panic or choose to shoot it out, jeopardizing the lives of both police and witnesses. Both police pursuit and investigation can be expedited if an employee from the victimized business is sent outside to inform the arriving officers that the robber has already left.

Interviews and Statements by Witnesses or Management

Management should make certain that the police are furnished a private place to interview individual witnesses. Employees who have just gone through a robbery need an opportunity to think calmly, to organize their thoughts without additional distractions.

Then, too, a witness may sometimes hesitate to furnish some information unless this can be done in confidence. Many persons are reluctant to tell what they know if they think the bandit, still at large, may be watching them on the ten o'clock news.

In some instances, management may allow TV reporters or newsmen with cameras to crowd into interviews. This almost always hurts the investigation and tends to slow down the opportunity for hot pursuit. In addition, reporters will seldom hesitate to broadcast information regarding leads that police have developed, even though this action may tip off a suspect as to where a trap has been set or what evidence should be destroyed linking the bandit to the crime.

Sometimes management hesitates to bar entrance to representatives of the news media, for fear unfavorable publicity will result. And newsmen have every right to go into public areas. But a businessman can lock the front door when he chooses, admitting the police and no one else. If employees are willing to be interviewed by newsmen, then it may be allowed after the police interviews. Management should not, however, order an employee to be interviewed by TV or press reporters if the employee does not want to go through this procedure. If a TV reporter should arrive on the scene before the police, there is a good possibility that independent descriptions of the bandit may be colored if witnesses are interviewed in the presence of others.

Statement to the Press About the Amount of Loss

It may often be questionable judgment for management to make a statement of the amount of money taken in a robbery. Sometimes this information will be available to reporters anyway, if they are given access to police reports.

If management's press release indicates that a large sum of cash was taken, then other criminals may be encouraged to rob the business a second time. If the release reflects that the bandits missed a quantity of money that was hidden in a back room, this may also motivate a second robbery. Criminals may assume that a large haul can be made by forcefully threatening the merchant to reveal the secret hiding place of these surplus funds.

SUMMARY

Robbery has long been a major cause of loss for some businesses, especially supermarkets, groceries, and small shops. Frequently committed on an impulse by dope addicts or criminals short of funds, robbery is not always a logical crime. Planning to avoid robbery is of definite value in reducing robberies by calculating professionals, however.

There are three basic security objectives in combating this problem:

1. Employee and bystander safety.
2. Identifying and apprehending the criminal.
3. Avoiding or recovering loss.

Because of the stress placed on employees by an actual robbery situation, it may be helpful to provide all employees with a wallet-sized instruction card, setting out step by step the action that should be taken in a holdup. Employees should also be trained to recognize unusual customer actions that may mean the business is being cased for a robbery. Silent alarms, CCTV installations and recorded bait money slipped in with the loot are all ways to combat robbery.

Kidnapping of a store or bank manager is a technique robbers sometimes use to gain access to a safe; the kidnappers force the manager to come to the business with them and open the safe while members of the criminal gang hold the manager's family at gunpoint in their home. Individuals responsible for large amounts of money must be sure never to let strangers into their homes.

Minimizing the loot available, by setting a cashier's cash limit, by bleeding cash registers, and by sending money to the bank or night depository is usually helpful in discouraging robbery.

Proper location of the cashier's facilities and money room can sometimes prevent robbery. Paying employees by check, and retaining incoming checks separate from cash may also help.

Opening-time holdups can usually be avoided if one employee waits outside to receive a signal from an entering employee that the building is free of intruders.

Available TV or camera systems and silent alarms should be set in action as soon as possible after a robbery beings. Giving a prearranged signal to co-workers that a robbery is in progress can also speed police response.

Obtaining an accurate description of each bandit, especially as to unusual features, is often essential to solving a robbery. Observing the getaway route and recording the license number of the getaway car may also be very helpful. Police authorities, not the company auditor, should be contacted immediately by telephone as the bandits depart, and the line should be kept open to the police so developments can be reported immediately. It is also helpful for an employee to go outside to meet arriving police, so that they do not waste time making a cautious entry to the premises.

Care should be taken to see that any place touched by the criminals, or any object that may be evidence, is protected. Names and telephone numbers of witnesses, along with their car license numbers, should be recorded immediately. Ideally, employees and other witnesses should be isolated from one another and given a sheet of paper on which to write down a detailed description of each bandit. This information may be of great value to law enforcement in solving the crime.

REVIEW AND DISCUSSION

1. Describe the basic differences between robbery and burglary.
2. What are the objectives that security representatives and police officers must bear in mind in preventing and solving robberies against business?
3. Explain how the increased use of narcotics has affected armed robbery.
4. What is a so-called robbery procedure instruction card? How is it used? What are some items that should be listed on an instruction card of this kind?
5. How can a clerk in a business recognize that the business may be being cased for a robbery?
6. How many business managers and officials avoid being kidnapped by armed robbers?
7. Why is it helpful to minimize the amount of loot? Explain.
8. Should checks and other business papers be protected? Why?
9. What may be done to prevent a robbery at the time employees open the business?
10. What should an employee do during the course of a robbery? Explain in detail.
11. How should the robber's description be recorded?
12. List the basic steps that should be taken after the robber leaves.

10

Business and Residential Burglary

This chapter considers the security problems associated with the crime of burglary. No one can prevent burglary absolutely, but it will be less frequent and losses will be reduced if good security measures are taken, Perimeter protection, fencing, security lighting, sturdy doors and windows, adequate locks, alarm systems, guards, and other security techniques are all helpful. In addition, the three basic types of burglars are described: the professional, the cat burglar who may be a narcotic addict, and the juvenile. Differences in the approach of these three types are examined. Finally, consideration is given to how burglary losses can be reduced and how burglaries may be solved.

BACKGROUND

What Is Burglary?

Legally, burglary is defined as a breaking and entering of a business or dwellinghouse for the purpose of committing a felony inside the building. Usually the felony intended inside the structure is theft, either of money or valuables. However, intent to commit any other felony inside the building also constitutes burglary.

As noted in the material on robbery, burglary is usually a crime against property, whereas robbery is a crime against the person of the victim. Sometimes burglary is a crime against both property and the person, as will be pointed out in the discussion of cat burglars. Frequently, burglary is a crime without witnesses, often provable in a criminal trial from the physical evidence left at the scene.

Security representatives should keep in mind that occasionally a reported burglary is false—a contrived situation, staged to cover an employee theft or some other internal crime.

It should also be realized that no security program can absolutely prevent burglary. But good security measures can create an environment in which burglaries are less likely to occur. This is the approach that police officers and experienced security professionals sometimes call hardening the target.

Conditions Behind Increases in Burglary

Within recent years, burglary has increased at an alarming rate, and this increase seems destined to continue. A number of factors contribute to this ever more serious problem:

1. Crime is generally concentrated where the population is most dense. Approximately 33 percent of all burglaries occur in the ten most heavily populated areas of this country.
2. Since more women are now employed, more homes and apartments are left unattended for extended periods.
3. There has been a marked increase in population growth in the age group most likely to be involved in burglary (males from 18 to 25 years of age). There is also more unemployment in this age group, and the unemployed frequently turn to crime.
4. Roughly 4.5 percent of the families in the United States now have incomes of approximately $25,000 per year, or more. These families are natural targets for burglars.
5. More and more American families are able to purchase second homes or vacation cottages. This makes unoccupied residences a target for looting or vandalism.

Three Basic Types of Burglars

Experienced police authorities on burglary sometimes classify burglars into three broad categories:

1. *The professional.* Usually this is an individual with the skill to enter a well-protected building, disarm an alarm system, and break into a safe.

The elite criminals in this category are sometimes called safe crackers or yeggs. In general, professionals are confirmed offenders who realize that they may do time in prison; they approach burglary as a profession with a calculated risk.

2. *The cat burglar.* This type of burglar frequently enters houses or apartments as well as businesses, usually armed with a gun or a knife. Many are dangerous drug addicts, seeking cash to support a costly narcotics habit. Police officials report that some cat burglars are mentally unstable. If these criminals are under the influence of narcotics when they commit their crimes, they may assault, kill, or rape if confronted. Whereas most burglars seek to avoid confrontation, some cat burglars seem almost to welcome coming face to face with the victim. Although cat burglars do become involved in business burglary, they usually concentrate on residential crime.

3. *The amateur.* This type of burglar is usually under 25 years of age. A high percentage of these individuals are from broken homes and are high school dropouts. They may be either black or white and frequently have a history of truancy, petty theft, and auto theft. Amateur burglars seldom have the skill necessary to break into a safe, but they will enter business buildings or stores regardless.

What Are Real Losses in Business Burglary?

Statistics maintained for a number of years by insurance associations indicate that the cash loss in the average commercial burglary may be less than $200 per incident. Additional statistics indicate that the loss from building and equipment damage and theft is often as great as the actual cash loss sustained in a business burglary. If doors are broken open or windows smashed, replacement costs may be considerable. In schools and institutions, juvenile burglars often engage in highly destructive vandalism while commiting a burglary.

If a professional burglar uses an acetylene cutting torch to open a safe, company ledgers, journals, and bookkeeping materials may be destroyed. Frequently, a building is set on fire by careless torch men. When nitroglycerin is used to blast open a strongbox, fire is sometimes a side effect of an improperly set charge.

Then, too, some burglars set a building on fire to cover up the crime. Even if the fire is extinguished by an automatic sprinkler system in the building, the fire damage to merchandise and building may be considerable. In addition, the water frequently damages retail goods, manufactured items, or raw materials. It is obvious, then, that the monetary and merchandise losses in a business burglary may represent only a part of the total loss.

THREE BASIC APPROACHES TO PREVENT BURGLARY

There are three basic approaches utilized in the security programs of most businesses that have successfully combated burglary. These are:

1. Reducing the time in which the burglar can work without being observed.
2. Reducing the amount of loot that is available in the event the burglar is successful in making an entry.
3. "Hardening the target" so that the burglar must go to great lengths to successfully penetrate the building.

Limiting the Time Available to the Burglar

Generally, burglars want to enter as quickly as possible, scoop up valuables, and get out of the building without delay or noise. If doorways and windows are in full view of the street, the burglar will realize that entry must be made in a minimum of time to prevent being observed by police patrols or passers-by. If an effective alarm system is in operation, the burlgar will realize that entry and theft must be accomplished with a minimum delay in order to escape prior to police response.

Almost any money safe may be openable by a top-grade professional burglar. However, a good safe can slow down the intruder for a long period of time.

If guards or security patrols are utilized, there may not be adequate time for entry between guard rounds.

If a company forklift is not available to the intruder, he may not have sufficient time to load small items of merchandise into a getaway truck. Also, if a thief must spend considerable time transporting merchandise to an open door or window, there is greater likelihood of apprehension while still in the building.

Limiting the Amount of Money or Loot That Is Available

As previously noted, cat burglars or amateurs may enter a business building at any time. Professional burglars, however, will seldom consider a business safe burglary unless there is reasonable assurance that a considerable amount of cash can be obtained.

In so far as practical, businesses should never accumulate large sums of money. Employees should make regular trips to the bank or use the services of an armored car delivery company. If possible to do so, the company security department should obtain management approval for a cash money limit and company tellers and cashiers should adhere closely to this policy.

Experience shows that it is usually advisable to let everyone know that the money limit is closely followed. If management insists on reducing the amount of loot that is available, this information usually becomes common knowledge in the criminal underworld. On the other hand, janitorial workers, repairmen, temporary laborers, and customers may be aware if the firm consistently violates regulations by exceeding the cash money limit.

Of course, if money is not readily available, burglars may remove electric calculators, typewriters, adding machines, or company merchandise. Items of unusual value such as watches, jewelry, and similar merchandise should be afforded additional protection inside the building. If an alarm system does not seem justified for an entire building, an alarm covering a special inside area may nevertheless be worthwhile.

The business can sharply limit the amount of loot the burglar can dispose of, if the security department insists on maintaining an inventory of mechanical equipment, typewriters, business machines, guns, and similar items by serial number. If these serial numbers can be furnished to the police immediately after a burglary, the crime may not be worth the risk to the intruder.

Hardening the Target

By hardening the target we simply mean making it physically difficult for any intruder to get into the building. For example, a first measure to take might be to erect proper fences or walls around the building to provide perimeter protection. This precaution may be followed up by installing and using night lighting, installing heavy doors, and protecting windows, skylights, coal chutes, and other openings. Obviously, good locks must also be installed and used. In some locations, burglary-resistant glass may be advisable.

Depending on the individual location and property values involved, it may also be advisable to use sophisticated access systems, alarms, and a well-supervised guard force.

These protective features and systems make the target very difficult for burglars to penetrate without being detected and apprehended. Any or all available physical techniques may be used to help harden the target. These applications and techniques are described individually in other sections of this book.

Laws Requiring Hardening of the Target: Leadership of Oakland, California, Police

For a number of years, available statistics have indicated that roughly three out of five forcible entries into business buildings were made through doorways. In looking for ways to cut down on this type of burglary, the Oakland, California, Police Department conducted studies that disclosed that more than 70 percent of business burglaries in that city in the early 1960s had occurred in insecure premises. Available facts indicated that the majority of these burglaries were perpetrated by opportunistic criminals, rather than by skilled professional burglars.

As a result of the leadership of the Oakland Police Department, an ordinance was passed into law in that city in 1963, requiring business property owners to maintain basic standards of physical security. This ordinance included criminal penalties that could be imposed on businessmen who persisted in making insecure premises. Revised regularly since 1963, the Oakland ordinance gave police the right to suggest and recommend improvements for businesses that repeatedly experienced unlawful entries.

The Oakland ordinance did not completely eliminate commercial burglary. Enforcement of the ordinance did, however, require businesses to provide adequate protection, with the result that reported burglary violations fell drastically.

The Oakland burglary and security code is regarded as a model, setting minimum standards and requirements. This code has been copied and utilized in other cities throughout the country. Other cities have copied this code in city ordinances to curb burglaries.

Protecting the Safe Combination and Building Keys

In a recent case at Downey, California, burglars obtained $17,000 in cash from the safe of a labor union, utilizing the safe combination that had been written out and left under a desk calendar by an employee. If possible, employees should memorize the safe combination or keep the combination in a locked area where it cannot be discovered through accident. In addition, it may be helpful to disguise the combination by writing fictitious digits before or after the numbers in the actual combination.

Some burglars gain access to business or residential keys when the victim parks in a commercial garage or parking lot. All keys except the vehicle ignition key should be retained by the person leaving the vehicle. If an entire ring of keys is left, all the keys may be duplicated in a few minutes by the parking lot attendant. From the automobile registration inside the glove compartment of the car, the parking lot attendant can usually obtain the home or business address of the person leaving the automobile. Equipped with the address of the victim and a ring of duplicate keys, a burglar is way ahead of the game.

Some Techniques Used by Professional Burglars

Some authorities on business safes have estimated that approximately 75 percent of the business safes in operation are obsolete. In other words, the locking devices and protective features of these safes are not good enough to resist the efforts of a good professional burglar.

There are wide differences among professional burglars with regard to skills and training. A number of different techniques may be used to open safes. Some burglars specialize in one particular type of entry, but techniques may vary depending on the construction features of the safe.

At times, professional burglars simply beat open a safe with a heavy sledge hammer. This usually takes considerable time and energy and thus is attempted only after the safe has been picked up with a forklift or rolled into the bed of a waiting truck and carried to a remote location.

In other instances, the entry may be made by ripping or peeling open the safe. This technique is frequently used to enter fire safes rather than money safes that are equipped with heavier steel plates. The burglars either insert a large pry bar with a sharp end into the crack between the safe door and the frame into which the door fits or force the sharp end of the bar through the outer steel plate. Once they insert the pry bar, they can peel or rip away the outer steel covering. The inner steel plate is then removed the same way. Better grades of safes can seldom be opened by this technique.

Well-constructed safes can sometimes be opened by a "punchman" who uses a sledge hammer to break the spindle off the dial of the safe and then drives the spindle into the interior of the safe by using a metal punch and a small sledge hammer. Some modern safes resist this type of entry, since they are equipped with a locking device that permanently freezes the locking bars when the spindle is knocked out. This arrangement in the interior of the safe is called a relocking feature.

Some burglars use a burning torch or acetylene torch to cut a hole through the side of a safe. Burglars using this technique usually need to have access to a water hose to cool the interior of the safe so that the contents do not burn. In some states, the use of the cutting torch became so common that special laws were passed setting unusually heavy prison sentences for individuals convicted of torch burglary. As a result, the incidence of such burglaries decreased in these states.

Burglars sometimes utilize nitroglycerine to blast open a safe door. This technique is effective only when the burglar is reasonably well trained in handling explosives and has access to nitroglycerine.

SOLVING BURGLARIES

Helping the Prosecutor Prove the Case

If a high percentage of burglary convictions are obtained in the courts, then burglars will tend to leave business alone. Some of these criminals may even consider it profitable to join the regular labor force.

In order to assist investigating police officers, businesses should maintain an inventory list of all items that may be stolen, especially typewriters, adding machines, calculators, hand guns, or any other items bearing a serial number. Marking property as belonging to the business is also helpful. This limits the thieves' available markets and often helps the prosecution obtain a conviction if police authorities recover identifiable stolen property from the burglars.

As previously noted, burglary is often a crime without witnesses. The chances for apprehending the criminal are greatly increased if the scene of the crime is left completely untouched until police investigators and laboratory technicians arrive. Microscopic evidence is sometimes vital and should be protected. All employees should be instructed to leave the scene completely undisturbed, especially the burglar's entry and exit points.

If a burglar has broken into a business safe, there is a possibility that microscopic particles of safe insulation have been caught in the burglar's pants cuffs or other items of clothing. This safe insulation can usually be examined microscopically and tied up to the safe that was broken open. If the burglar uses an acetylene torch to cut open a well-constructed safe, there is always a good chance that minute particles of metal from the safe itself or from the cutting torch have been sprayed into the burglar's clothing. A good crime laboratory can also frequently identify tool marks left by the burglar's tools, provided the tool marks are protected at the scene.

Masks, gloves, and other apparel may be helpful in identifying the culprit. Fingerprints and footprints also are frequently left at the scene. In addition, soil samples from areas around the building can sometimes be identified with soil in the burlgar's automobile or shoes.

SUMMARY

No security program can absolutely prevent burglary. But adequate measures can create an environment in which it is not likely to occur. This approach, known as hardening of the target, makes use of good locks, adequate doors, protected windows, guard services, alarms, and all the techniques or tools of physical security.

Some authorities on burglary classify burglars into three broad categories: (1) the calculating, technically able professional, usually in search of money; (2) the so-called cat burglar, often a narcotic addict armed with a gun or a knife and highly dangerous; and (3) the amateur burglar, usually young and inexperienced with little technical skill, often wandering aimlessly looking for anything that can be carried away.

Three basic approaches are usually taken to prevent or reduce burglary losses:

1. Reducing the time available to the burglar.
2. Limiting the amount of available loot.
3. Hardening the target to make the burglar do a great deal of work to be successful.

Safe combinations should be regularly changed and protected and keys of all kind should be secured so that crime is not easy.

Because burglary usually is a crime without witnesses, it is especially important to leave the scene of the crime undisturbed and to protect physical evidence or anything which could possibly be useful to police authorities.

Professional burglars may use a variety of techniques in opening a money safe. Sometimes they beat a safe open with a heavy sledge hammer. Sometimes they use a large pointed bar to rip or peel away the outer steel walls of a strongbox. A well-constructed safe may be opened by a "punchman" who uses a sledge to break off the spindle of the dial, thereafter driving the spindle into the interior of the safe with a heavy steel punch. Other entries into safes may be made by use of an acetylene cutting torch or a burning bar, or by use of a controlled blast of nitroglycerine.

REVIEW AND DISCUSSION

1. How does burglary differ from robbery?
2. Give a brief description of the characteristics of three different types of burglars.

3. What are the real losses in business burglary? Why?
4. How may the burglar's available time be limited or shortened?
5. What steps can be taken to limit the amount of money or valuables that may be available to the burglar?
6. What is meant by hardening the target?
7. What steps can security representatives take to help the prosecutor prove a burglary case in court?
8. Describe the techniques used by two different types of professional burglars.
9. Why should security representatives make certain that a burglary scene is protected until police officers arrive?

11

Merchandise Handling— Receiving, Shipping, Warehousing

This chapter examines some of the security problems encountered in obtaining, storing, and delivering or transferring merchandise. Special attention is given to dock problems and to problems relating to receiving of goods, since carelessness or falsification in this area is responsible for a great many thefts. Good stockkeeping and warehousing practices are discussed as well as methods for the safe delivery and transfer of merchandise.

DOCK PROBLEMS

Freight in motion is usually less prone to theft or pilferage than is freight left in staging areas, on docks, or in freight terminals. Such merchandise may be a prime target for thieves.

Dock thefts usually take place while employees are working inside the warehouse. But thieves sometimes are so bold as to openly grab articles while employees are actually working on the dock. Losses of this kind often can be prevented by stationing a uniformed guard on the outside dock until the merchandise is brought into the building.

If freight is received only at specified hours, it is easier to see that the dock will not be left unattended. If it is at all possible, shipments should not be left on the dock while all employees go for coffee or lunch breaks. Usually, break periods can be staggered so that some employees are on hand at all times.

Ideally, warehouse doors should be closed at lunch or break time. This is frequently impractical, as warehouse temperatures may be so high that ventilation through the doors is an absolute necessity. The cost of lockable, woven-wire gates is almost always justified, provided the gates are of sufficient height, are locked in place, and the wire on the gates does not allow articles to be passed through.

If an unusually large shipment is received, there may not be enough warehousemen on duty to bring all merchandise into the warehouse or stockroom. Workers should be encouraged to bring in first those items most likely to be stolen. Anything left on the dock may, of course, be tempting to a thief, but some items may be too heavy to carry, so it is preferable to bring in items of small size and great value as soon as possible. Employees should be trained to be selective in bringing in freight.

Insofar as possible, only delivery or shipping vehicles should be allowed in the receiving area. It is also advisable to locate employee and visitor parking lots at some distance from both receiving and shipping docks.

Case after case has shown that a dishonest delivery driver may load merchandise back into the delivery truck after it has been counted by the receiving clerk. Experience has also shown that the delivery driver may place one or two boxes in the cab of the truck prior to arrival at the dock, hoping that the load will not be counted accurately and it may be possible to retain items left in the cab.

It is generally advisable for receiving and shipping to be handled on two separate docks, but at times, the layout of the company building will not permit construction of separate docks. When one dock is used simultaneously for shipping and receiving, there is increased likelihood that a carton from an incoming shipment may be carried away by a departing driver.

If the same dock must be used for both receiving and shipping, it may be possible to schedule the two activities at different hours. A substantial dock divider can be constructed by welding heavy wire mesh on a pipe frame, with wheels at both ends of the divider. This portable divider can then be moved up and down the dock to adjust dock space according to need, with shipping and receiving kept separate.

Company managers and warehouse superintendents sometimes state that it is practically impossible to restrict access to the interior of the company warehouse. But they have been proved wrong in business after business where rules have been conscientiously followed. Experience shows that losses are likely when truck drivers, customers, or sales employees are allowed to wander in the stockroom or warehouse areas without control. It is unquestioned that the job of escorting outsiders places a heavy burden on warehouse employees. But inventory shrinkage figures usually prove that access to the warehouse should be granted on the basis of need only. Outside delivery drivers usually need access to rest rooms, a public telephone, vending machines, and a drinking fountain. If drivers must wait for long delays, they also may need a lounge. In planning warehouse accommodations, management should see to it that these necessary facilities are located immediately adjacent to the docks where drivers or deliverymen are loading or unloading.

Experience in the shipping industry also shows that drivers should be clearly told that they are restricted to certain areas. Some companies will not allow a driver to continue to bring freight to their warehouses unless company regulations are followed.

RECEIVING

Businesses sometimes find that poor merchandise receiving controls may contribute as heavily to employee theft as poor shipping controls. Sometimes a firm has good controls on shipments received from outside suppliers but no controls on incoming merchandise from other stores or warehouses owned by the same company. This failure is sometimes explained by the statement, "Our shipping employees are always accurate, so recounting would be only a duplication of effort."

Experience shows that if it becomes known that an incoming shipment from another warehouse or store is not being counted, the employee forwarding the merchandise may ship only part of the order and steal the remainder.

In a recent case involving a Dallas drug warehouse, there were no controls between the main warehouse and a backup warehouse used to handle overflow shipments. When the main warehouse ran short of merchandise, a truck driver ws sent to the backup warehouse to pick up whatever was needed. Since there were no controls on receiving between the two warehouses, the truck driver habitually picked up an excess of merchandise worth from $3000 to $10,000 on each trip. He then stockpiled this excess merchandise in his garage, since he lived along the route usually followed between the two warehouses. After a time the stolen drugs, which had cost the drug wholesaler approximately $80,000, were sold on the black market by the guilty driver.

It is also suggested that company managers and supervisors should not receive personal shipments or deliveries at the company office, unless these deliveries go through the regular receiving process in the warehouse. If owners or managers of the business object to having their personal packages opened, then these deliveries should be made to their home address. No incoming packages should be exempt from the receiving process.

Receiving clerks often point out that it is difficult to count accurately in a confused work area. To avoid distractions, some firms permit receiving only at specified hours. These companies often go so far as to decline to receive shipments that arrive at other hours.

At retail stores, merchandise should be received through a rear door or a side door of the building, if this is possible. In some metropolitan areas, however, the alleyways may be so congested that incoming freight must be left on the sidewalk. This generally interferes with customer access to the building and passers-by may be injured as a result of this practice.

Receiving "Blind" as a Good System

A receiving clerk will sometimes fail to count incoming merchandise unless required to do so by the system that is used.

At the time merchandise is purchased, some firms send two copies of the purchase order to the warehouse office or to the warehouse superintendent. One copy of this purchase order, which is usually given to the receiving clerk, does not contain information as to the quantity of merchandise that has been ordered. This is because this copy of the purchase order has a special process in the paper that does not permit quantities to be read. The receiving clerk therefore knows what kind of merchandise is expected and the name of the shipper, but is forced to count the quantities when the merchandise arrives.

In a system of this kind, the receiving clerk must make an independent count of the merchandise and fill out a blank receiving report form. The second copy of the purchase order in the warehouse is maintained in the warehouse superintendent's office and this copy reflects the quantities ordered. After the receiving report has been prepared by the receiving clerk, the warehouse superintendent compares the receiving report with the purchase order that reflects the correct number of items ordered. An immediate recount is made if the quantities on the receiving report do not correspond to the quantities on the purchase order.

In counting incoming freight, a portable lighting system is sometimes helpful. It may be desirable to utilize a floodlight that can be moved from trailer to trailer or truck to truck as compartments are being unloaded.

As a part of the receiving function, the receiving clerk must also determine whether items received are the right kind of merchandise, or whether a cheaper item has been substituted. Losses result if the receiving clerk does not notice the substitution of an inferior product.

Receiving clerks sometimes make a practice of receiving from the packing slip that is inside the incoming merchandise box. Often this packing slip is very hard to read and errors occur if the packing slip is accepted at face value. In this kind of receiving, the clerk is, in effect, accepting whatever errors may have been made by the shipper.

There is an additional administrative problem created when a "blind" receiving system is used. If the shipment is deliberately short, the receiving clerk who is counting "blind" will not be aware of that fact. The receiving clerk turns in the count that he actually found to be accurate. This count is then given to the warehouse superintendent or the individual in charge of receiving, who compares it with the purchase order. A notation is then made on the second copy of the purchase order which is maintained in the shipping section, noting that the shipment was only partially received. A copy of the purchase order is then maintained in a suspense file, awaiting arrival of the remainder of the shipment. When the balance of the merchandise is received, the suspense copy is marked as received and sent to the business office for payment.

Security Department Verification of Receiving

From time to time, the company security representative should make unscheduled checks of items received and counted in by the receiving department. When a verification of this kind is anticipated, receiving clerks are far more apt to count in a conscientious manner.

Receiving Overshipments

Some companies refuse to accept overshipments because they may not be able to dispose of the extra merchandise immediately and have no stockroom or warehouse space in which to store it. If the receiving clerk discovers an overshipment after signing for the regular shipment, there is always a possibility of theft of the excess part of the shipment. The receiving clerk may put the merchandise in stock, hoping that an accounting will not be required for the extra merchandise. Subsequently, the receiving clerk will take the extra merchandise out to his personal car or ship it through a cooperative truck driver.

SHIPPING

If a company form—sales ticket, invoice, etc.—is used to move merchandise from one place to another, then these forms should be numbered and accounted for by number. To avoid theft or embezzlement, someone must check to see that there is billing or receipt of money for each form issued.

Businesses sometimes relax controls before the Christmas season, permitting company employees to ship Christmas packages by using company facilities. The only expense involved may be the time devoted to shipping these packages by the shipping clerk. But it is sometimes found that employees include some company merchandise in their shipments. Workers may also waste valuable time if they are allowed to wrap gift boxes while on duty.

Shipping by Mail or by United Parcel Service (UPS)

Many companies make shipments by Parcel Post or United Parcel Service. The UPS meter issues a strip with a specific amount of money representing the cost of the shipment. The meter registers the exact number of stamps (strips) that have been issued. A system should be used to count the number of stamps issued and to compare this count with the number of packages or boxes awaiting pickup by the UPS truck driver. Both of these figures should also balance with the number of shipping orders that have been received for UPS shipment on that date. If the check shows that more boxes are awaiting UPS pickup than stamps have been issued, the extra box may be one containing stolen merchandise that is being shipped to an employee's home. The usual method used by a thief in this situation is to run off an extra strip stamp and to hold this strip stamp until the stolen merchandise can be packaged for shipment. Usually the thief waits one or two days to use the extra stamp.

To maintain security, it is also important for the UPS meter or postage meter to be locked up at the close of business. The key to the meter should be kept in a safe place or the meter itself should be locked away.

Thefts from the Delivery Truck

Some companies experience losses from the tailgate or rear of delivery trucks when the vehicle stops at a traffic light or to make a delivery. A thief may simply jump onto the tailgate, grab one box, and jump off before the driver starts again.

Freight is frequently stolen from local pickup and delivery trucks simply because no padlocks are supplied to the driver. Sometimes locks will have been furnished, but they remain attached to the rear door of the units involved without ever being used.

In case of engine trouble, the driver should have padlocks that can be applied to all doors to make certain that merchandise is secured prior to the time that the driver goes for help.

Shipping of small items of great value has long posed a security problem. Some companies have found that packaging portable, high-value items together in one package of considerable size and weight is helpful. If the package is large enough and heavy enough, thieves may not be able to move it without special equipment.

This procedure is recommended especially for shipments of items such as diamonds, gold, and other high-value merchandise. Sometimes a shipping agent places special steel or concrete "igloos" over pallets containing high-value shipments. Gold bars may be placed in heavy steel boxes that are first welded shut and thereafter welded to a heavy steel pallet. Some shippers have gone so far as to weld heavy steel boxes containing gold bars to the bed of the transport truck.

Theft of the Entire Load

If a trucking company does not take security precautions, thieves may steal an entire truck or trailer containing valuables. Once the theft has taken place, the criminals responsible can unload the shipment at their own convenience. Experience shows that in theft situations of this kind, the truck driver is often a conspirator with the outside thieves.

The following precautions can be taken to prevent theft of an entire trailer or truck loaded with merchandise:

1. Nonstop hauls should be made, where possible. Time is of the essence in moving high-value cargos through high-risk neighborhoods.
2. Parked trailers or trucks that contain valuable merchandise should be left under good spotlights in freight terminals or parking areas.
3. Loaded trailers should be parked back to back in terminal yards or other parking areas so that the rear doors cannot be opened without attaching a tractor to one of the trailers and separating the trailers. This preventive method does not work if the trailers can be opened from side doors.
4. Trucks should move in convoys if they are going to the same destination loaded with valuable cargo. In some instances, it may be worth the expense to have a company security agent or supervisor follow a vehicle in a private automobile.

5. Painting the company name and truck or trailer number on the top of the vehicle is often helpful. Most state police agencies now use helicopters to look for stolen vehicles or trailers and the marking on the top can be very helpful, provided the lettering is large enough to be read from the air. Painting the company insignia on the sides of the truck may also be a deterrent in many instances.
6. If the trailer contains merchandise of unusual value, and this fact is known at the shipping terminal, it may be desirable to cover the number of the trailer temporarily.
7. Teletype or telephone messages about the arrival and departure of value shipments should be closely guarded. Departure times should be known only to those employees who are immediately concerned with the shipment.
8. If a loaded trailer must be left in a freight yard or delivery terminal, a kingpin lock should be placed on the kingpin or fifth wheel to prevent an unauthorized tractor from being backed up to the trailer to pull the trailer away. These locks are well-constructed, machined from solid steel, and very difficult to remove.

Preloading of Delivery Vehicles

Loading of a delivery truck is often a time-consuming process. Merchandise must be loaded during nighttime hours or on the day prior to delivery. This preloading process may create security problems that do not otherwise exist. Careful supervision usually limits theft, but if preloading is done at night, supervision is often at a minimum.

Recently, the warehouse supervisor in a St. Louis company had been instructed to verify all merchandise placed on outgoing vehicles. However, because of numerous other duties, the supervisor checked only the loads of newly hired drivers. One experienced driver with a long record of honest service observed what was taking place. This driver began to load color TV sets on his own delivery vehicle as soon as he was reasonably sure that his load would not be verified.

Some companies have found it preferable to preload delivery vehicles on the afternoon shift, as soon as the trucks return from their daily runs. This practice solves the problem of inadequate night supervision and keeps the company from having to pay overtime for nighttime labor.

Some companies leave preloaded vehicles inside a locked, fenced yard overnight. Individual alarm systems on the parked trucks may be justified in such a case. It is also recommended that each door of each vehicle be equipped with a separate padlock and that the locks be applied at the time loading is completed. Keys to these locks should be controlled so that only the delivery driver and company supervisors have access to the keys. Obviously, ignition keys should not be left in the vehicle.

An industrial linen supplier in Los Angeles made a practice of loading delivery trucks and leaving them parked in a well-fenced yard. Knowing that keys were always left in these vehicles, an ex-employee with a grievance climbed over the fence one night and drove one of the loaded trucks through the fence. He kept on driving, throwing out part of the load as he went, and finally dumped the remainder into an alleyway just before abandoning the truck.

Even when ignition keys have been removed from preloaded trucks, however, the trucks may be vulnerable. A wholesale grocery company in the Des Moines area regularly left locked, preloaded vehicles on a fenced company lot. The keys to these vehicles were maintained on a board located inside the company guardhouse at the warehouse entrance.

To obtain vehicle keys, truck drivers had to report to the guardhouse and present company employee passes. The guard on duty would then verify the authenticity of the pass and allow the driver to pick up keys for the appropriate vehicle.

When the guard was distracted, one driver picked up keys for his own vehicle as well as for another vehicle that was being repaired. The driver then duplicated this second set of keys and returned them before they were missed. When the truck to which these keys belonged was back in service, this driver furnished the duplicated keys to a confirmed criminal who used them to drive the truck away with a load of goods worth approximately $80,000.

Some firms allow delivery drivers to take a preloaded truck home. The driver parks the vehicle at the driver's residence during the night. Companies that have used this system point out its advantages, but there may also be security weaknesses in a system of this kind. Individuals in the neighborhood can usually tell when the driver and members of his family are away from home. Thieves may take advantage of such an opportunity to loot the truck.

Surveillance (Tailing) of Company Trucks

Management has a legal right to conduct a surveillance of, or to tail, company vehicles. Such occasional surveillance will frequently reveal some helpful facts about customers and about the activities of drivers. Companies sometimes find that a driver is making unauthorized stops at retail customers. These stops may be well-intended, or they may be instances in which the driver takes merchandise off the truck and sells it without authority.

On occasion, it may be observed that the vehicle driver has very poor driving habits and is thus an insurance risk, or that the driver forgets to remove ignition keys when making a delivery, leaving the vehicle and the remainder of the load to be easily driven away by a thief.

On the whole, delivery drivers are conscientious, loyal employees. But occasionally a well-conducted surveillance may reveal that a driver is picking up hitchhikers against company rules or spending large amounts of time drinking coffee or intoxicants with company drivers or drivers from competing firms.

Using Railroad Seals to Control Shipments

Many shipping firms use railroad seals to verify the integrity of the shipment. These seals are thin strips of metal fashioned into a sealed loop with a crimping tool. Although these strips can be removed and then sealed again by crimping, this can seldom be done without leaving very noticeable marks. Some companies place the firm name on all seals and the seals are numerically controlled. Most companies that place the seal on a trailer, railroad freight car, or shipping container record the number of the seal on the shipping papers. The individual receiving the shipment can then compare the number on the seal with the number recorded on the shipping papers. The seal record, then, provides a reasonable assurance that the shipment has not been opened while en route. Unless the receiver makes a careful comparison of the seal numbers with these noted on the shipping papers, there is always a possibility that a thief may have removed the original seal, stolen merchandise, and placed a second seal on the shipment.

If a sealed delivery truck is to make a number of stops, the truck driver should be furnished with additional seals for each stop made. When a seal is broken and merchandise removed, the driver can then reseal the load until the next destination is reached.

If a receiver finds that the seal numbers on a shipment do not correspond with those on the shipping papers, it is usually advisable for the receiver to immediately contact the responsible officials at the shipper's headquarters. If this action is taken prior to actually entering the sealed container, there should be no legal dispute about which company is responsible for any missing merchandise.

WAREHOUSING

Selecting and Checking Shipments

Thefts frequently occur in a warehouse or stockroom because controls on transfer of merchandise are inadequate. Warehousemen or truck drivers are able to load merchandise into a delivery vehicle or to ship it without being held accountable.

Usually, sales tickets originate in a company office. Copies of the sales ticket are then sent to the warehouse to be filled and shipped.

A method some companies use to reduce opportunities for theft at the warehouse is to have an employee called a selector or picker select the merchandise called for by the sale order. The merchandise then goes to a second employee who verifies the accuracy of the selected merchandise. This second employee is called a checker or a quality control employee. Sometimes the merchandise goes to a third employee, a packer, after being checked. The packer wraps the merchandise and puts the appropriate address label on the outgoing box or boxes.

Some companies combine the functions of the checker and the packer. Still other warehouse systems allow one employee to perform the duties of the selector, the checker, and the packer.

For good security, an independent count of the outgoing merchandise should be made by the checker and this individual should be a second employee rather than using the selector for both functions.

Often companies find that warehouse thefts are reduced significantly if a representative of the security department or a member of management verifies the merchandise in outgoing shipments, checking the shipment just before it is wrapped by the packer. Another effective technique is to wait until a departing truck is loaded and then take off the load and compare the merchandise on the truck with the shipping documents reflecting the merchandise that should be on the truck. If more merchandise is found than can be accounted for, there is a strong indication of theft.

Some firms have reduced warehouse and shipping thefts by requiring the warehouse superintendent to observe loading, at least on a part-time basis, and to make unannounced checks of individual shipments going onto the truck.

Most experienced security representatives are convinced that the truck driver should not be allowed to assist in the loading, unless the company does not have sufficient personnel for this job. A high percentage of thefts in shipping by truck involve a dishonest truck driver working with a loader who puts on more merchandise than required.

A security technique sometimes used is "salting the load." What happens here is that someone from security or management loads an extra piece of merchandise on the truck, without the knowledge of the driver. If the driver is dishonest, the surplus item may not be turned in when the driver comes back to the warehouse. This technique is seldom effective if used frequently because drivers are quick to catch on that the company is playing a game.

Controls Over Returned Merchandise

Procedures for merchandise delivery should include controls over returned merchandise. Warehouse employees sometimes take the attitude that goods returned for credit belong to no one in particular and thus may be taken by anyone. Returned goods should be promptly returned to stock rather than retained in a separate area in the warehouse.

Experience also shows that a company delivery man sometimes issues false credits to an outside business. It is wise to require verification before giving credit for returned goods. A supervisor should be on hand to observe the actual return of the merchandise and to record the reason for its refusal. When possible, this should be done on a daily basis and the merchandise returned to stock as soon as possible.

SUMMARY

Experience shows that freight is far less likely to be stolen if it can be brought into the warehouse from dock areas as soon as it arrives. Experience also shows that losses are likely to increase if shipping and receiving are conducted simultaneously on the same freight dock. Some solve this problem by shipping at specified hours and receiving at others. Portable dock dividers that can be adjusted to meet space needs are helpful, but maximum security is afforded by having separate docks for shipping and loading in different areas of the warehouse.

If outside truck drivers are restricted and controlled, better security usually results. Thefts from warehouses decrease when company salesmen, customers, and other unauthorized individuals are not allowed in the warehouse unless escorted.

Heavy wire security gates locked across warehouse doors will almost always justify their cost by keeping out intruders while admitting needed ventilation.

Some businesses report that poor merchandise receiving controls encourage employee theft as greatly as do poor shipping controls. If an employee responsible for forwarding merchandise knows that incoming shipments are not being counted properly, he may forward only part of an order and keep the remainder. Shipments between two warehouses or stockrooms in the same company are especially susceptible to theft if employees forego the use of regular receiving procedures. In general, most authorities on security in this field believe that the best systems are those that require the receiving clerk to receive "blind," that is, without knowing in advance the quantities of merchandise in a shipment. For this system to be effective, the receiving clerk must fill out a separate written receiving form, and the warehouse superintendent or another employee must compare this receiving form with the purchase order to determine if the shipment is short or over.

To avoid thefts in the shipping process, many firms use numerically controlled shipping or sales tickets, accounting for each on a daily basis. If a management or security representative makes a practice of occasionally verifying the contents of an outgoing shipment, the likelihood of a warehouse thief shipping out more than has been paid for is significantly decreased. It is also helpful for the warehouse superintendent to supervise the loading of each shipment and to keep the truck driver from loading his own truck. If time permits, a good security measure is occasionally to unload an outgoing truck, comparing the merchandise on the vehicle with the items called for on the shipping ticket. This method is especially useful in detecting thefts if employees have no advance knowledge that the truck is to be unloaded.

Deliberately placing extra merchandise on the truck (salting the load) is a technique that can also be used to detect a delivery driver who is dishonest.

Unless a delivery truck en route is locked, a thief may jump on the rear of the vehicle and make a tailgate theft. If the truck breaks down and the driver must go for help, he should have padlocks with which to secure the load. A driver should never leave the keys in the ignition when he goes in to make a delivery.

Sometimes thieves steal an entire loaded trailer or truck; to prevent such a theft, a company can require nonstop hauls or convoys, or assign a security agent to trail the shipment. Loaded trailers should be left under good yard lighting, parked back to back, and the numbers of the trailers should be covered while they remain in the freight yard. Painting the company name on top of a vehicle makes it easy for a police helicopter to spot. Kingpin locks are recommended for securing preloaded trailers. Teletype or telephone messages about arrival and departure times of value shipments should be kept confidential. Preloaded trucks are vulnerable to theft.

Management has a legal right to tail company delivery trucks. This procedure sometimes reveals that a driver is dropping off stolen merchandise during a scheduled run.

Railroad-type seals should not be relied on to secure truck doors or railroad cars, but they do alert management to the fact that someone has broken into the sealed shipping container.

REVIEW AND DISCUSSION

1. Why is it important to bring incoming merchandise inside the warehouse as soon as possible? Explain.
2. Should warehouse doors be closed when employees leave for lunch or coffee breaks? What are two ways of maintaining security during times when employees are likely to be absent? Describe what may be done here.
3. Is it advisable to set up separate shipping and receiving doors in the warehouse or stockroom? Why? If this is necessary, how may it be done?
4. Is it always necessary to maintain controls over shipping between company warehouses or stores? Why?
5. Explain what is meant by receiving "blind." Describe advantages and drawbacks to a receiving system of this kind.
6. What can be accomplished by a verification of receiving by a security department representative or a member of management?
7. Explain how a thief can use United Parcel Service (UPS) or Parcel Post facilities to ship stolen items out of the warehouse or stockroom.
8. What is to be accomplished by checking outgoing shipments or supervising the loading of outgoing shipments?
9. What can be done to prevent thefts from the company delivery truck?
10. What steps may be taken to prevent theft of an entire truck and cargo, both when the truck is parked and when it is in transit?
11. List some ways of protecting a parked trailer or truck loaded with merchandise.
12. Outline some of the risks involved in preloading trucks for delivery on the following day.

13. With regard to security and safety, what may be learned from a surveillance of a delivery truck?
14. How are railroad seals used to protect freight shipments in transit? Outline some of the precautions that should be followed in using railroad seals for security.

12

Personnel Security

In the final analysis, all businesses must reply on the honesty of individual employees. The best physical security devices and procedural control systems may prove ineffective if honest employees are not available to make these devices and procedures work.

The purpose of this chapter is to outline the steps businesses should take to hire honest, dependable employees and how to make certain these workers continue to be honest and contribute to the security of the business.

OBTAINING HONEST EMPLOYEES

To restate the basic principle of personnel security, attitudes and honesty of rank and file employees are key to minimizing losses through theft. This does not mean that the effectiveness of physical and procedural controls is completely dependent on employees' cooperation but it is true that the advantages of such controls are significantly reduced if only unreliable individuals are available to lock doors, to put away cash receipts at the close of business, or to verify the accuracy of shipments received from an outside supplier.

Physical security devices and security systems can be changed merely by management decision. But a complete turnover of personnel is usually a time-consuming, inefficient operation. Then, too, productivity almost always suffers when experienced employees are replaced or discharged. It is therefore important to hire honest employees at the outset and to make certain that they remain honest. This should not be interpreted to mean that an ex-convict or a person with a doubtful background cannot become an acceptable employee. But a company's hiring policy must consider that experience shows a person will perform in the future much as that person has performed in the past. There are, of course, exceptions to this rule and no one should be denied the right to earn a living; but neither should a business be compelled to hire an individual for a sensitive job if the person has evidenced untrustworthy traits.

The company personnel official is often concerned only with the applicant's potential for job performance and for eventual advancement. But neither obvious ability nor latent talent in any applicant should influence the personnel interviewer to ignore factors that point toward the applicant's potential dishonesty or instability.

Interviewing Carefully

There are a number of techniques that may be used in selecting honest employees. If the company interviewer will carefully check former employees and references and fellow workers, questionable job seekers will often be eliminated. Some may even be weeded out at the time of the original interview. Rejection of an individual for a job does not mean that the applicant has necessarily been a bad citizen or a criminal in the usual sense of the word. But unexplained problems in the applicant's past job history may be revealed. Severe pressures in financial, marital, or personal areas often reflect a lack of basic stability or dependability that is needed.

A conscientious personnel interviewer watches for signs of drinking, marital or financial problems, habitual absenteeism, frequent job switching, or other possible indications that the applicant is unstable or immature.

A marital breakup is frequently an indicator of instability. But the fact that an individual has been married more than once should not serve to disqualify. At the same time, a lack of honesty in the applicant's attitudes toward marital responsibilities may indicate character traits that are not desired by the employer.

Insofar as possible, applicants should not be hired if the interview or the subsequent investigation reveals that they live beyond their resources. Should an unexpected need for money arise, such an employee may resort to theft to solve personal problems.

If the job applicant has been involved in bankruptcy or wage garnishments or has run up unnecessary debts, these are other indicators that job seeker lacks the dependability that usually goes hand-in-hand with employee honesty.

In addition, frequent address changes sometimes indicate an applicant's instability. In recent years, however, changing social and economic conditions have caused people to move with greater frequency than in the past. But it should be kept in mind that people sometimes move because they cannot pay the rent or want to avoid bill collectors.

Federal laws, designed to eliminate discrimination, have sometimes restricted the questions that can be asked a job applicant by business personnel interviewers. As usually interpreted, the questions that cannot be asked include the following:

1. Marital status—married, divorced, separated, widowed, or single.
2. Maiden name or father's surname, if the applicant is a woman.
3. The identity of persons who live with the applicant.
4. The church attended, or the name of the applicant's spiritual leader.
5. The number of children in the family, their ages, and information as to who will care for them while the applicant is at work.
6. Whether the applicant owns or rents a residence.
7. Whether the applicant's wages have ever been garnished.
8. Whether the applicant has ever been arrested. (The applicant can be asked to list convictions for criminal violations, if this information is pertinent to the job in question.)

Investigations of Job Applicants

Many employers believe that the most effective way to obtain good background information about a job applicant is to hire an outside investigator or to use a company investigator to contact all former employers, references, and a sampling of neighbors. Some firms reduce costs by conducting all or part of this investigation on the telephone, or by sending out form letters. Personal contact is almost always more effective in developing pertinent information, since many individuals are afraid to talk to strangers on the telephone or to write letters that could possibly be used in a lawsuit.

For many years, investigations of applicants by private or company investigators usually included a confidential check of police records. This was often the most effective way of determining whether the applicant had ever been arrested or convicted on any serious criminal charge. Today, police records are available to private investigators in only a very few localities and it is likely that records of this kind will be unavailable everywhere soon.

A check of credit bureau records frequently reveals whether the applicant is able to live on earned income, whether there have been lawsuits against the applicant, and whether bills are paid promptly.

The Polygraph; Psychological Stress Evaluator (PSE)

A detailed investigation of an applicant's background may be both costly and time-consuming. Businesses sometimes need to replace a key employee almost immediately and also want to obtain background information on applicants without spending a great deal of money. Feeling that conventional background investigations are an unaffordable luxury, some businesses utilize the polygraph as an alternative.

The polygraph, a device that is sometimes known as a "lie detector," is not a device for detecting lies. It is, rather, a testing device used to verify the truth.

In simplest terms, a polygraph test is based on the idea that lying leads to an emotional conflict that causes anxiety and fear. These emotions are reflected in physical changes in the subject's body. The polygraph records any changes in the subject's blood pressure, perspiration, and breathing (respiration). The principal objection made to the use of the lie detector in business or industry is that the test involves an invasion of privacy. Consequently, no test can ever be given without the complete consent of the individual involved, which is traditionally recorded in writing.

Historically, the idea for the development of the modern-day polygraph began with experiments to record changes in blood pressure that were noted by the Italian criminologist, Caesar Lombroso. Other Italian scientists, Musso and Benussi, experimented on recording changes in respiration while subjects told untruths. Leonard Keeler, a young American psychologist, refined the equipment and techniques that are used in today's polygraph.

The criminal courts have consistently refused to admit polygraph examinations as evidence in almost all criminal cases. The polygraph test is not a science, as definite as ballistics or fingerprinting. In the hands of a competent, experienced operator, the polygraph can be a highly effective tool. It should be kept in mind, however, that the accuracy of the polygraph depends on the skill of its operator.

The psychological stress evaluator (PSE) is a machine which detects, measures, and graphically displays some stress-related components of the human voice. It has been described as a cross between an FM radio receiver and a lie detector.

The basic idea of the PSE is that the human voice will indicate stress if the person being tested is undergoing anxiety, guilt, or fear reactions. The machine operates on the idea that inaudible frequency modulations are superimposed on audible voice frequencies to record the degree of stress in the voice of the speaker at the moment of utterances.

The psychological stress evaluator is a more recent invention than the polygraph. Some employers utilize PSE equipment to record all preemployment interviews on cassettes or reel-to-reel tapes. These recordings are then sent to an agency that is equipped to interpret the voice modulations.

As with the polygraph, written consent is usually obtained from the job applicant prior to the use of this kind of equipment.

KEEPING EMPLOYEES HONEST

Motivating Employees for Continued Honesty

An effective security program seeks to motivate employees to remain honest. Rank and file employees need to realize that failure to observe security regulations by one individual may create losses that jeopardize the jobs of everyone.

Experience shows that employees are usually unconcerned as to whether their employer is showing a profit or a loss. The attitude frequently expressed is, "This company has millions of dollars in assets; I am merely a low-level employee and there is no reason why I should become involved or be concerned."

But there are few firms so large that they may not fail. Many employees do not understand that their attitudes and activities affect company performance as a whole. Most workers think of themselves as completely independent of the company by which they are hired. The gains or losses attributable to one individual employee may not represent a large figure, but the company's profit or loss is nothing more than a combination of the efforts of all the individuals who work for the firm.

It is therefore important to motivate employees toward maintaining good security and to let them know that the company is relying on each one of them. Stated in other words, a security program requires individual concern by every employee who wants the company to make a profit. Unless profits are made, individual employees may have no opportunity for raises or advancement. And if the business continues to suffer setbacks, employees may find themselves out of work because of business failure.

It is important for managers and supervisors to create and maintain an atmosphere in which honesty is respected and in which workers want to do a good job. Almost all employees regard themselves as honest at the time of hire. Workers will generally accept high performance standards if these standards are presented in a proper way.

Experience shows that it is important for supervisors to assume that workers are honest. It is equally important for management and supervision to be scrupulously honest in their own attitudes toward outside customers, toward employees, and toward the company. This is because worker attitudes are frequently influenced by example. Experience also indicates that constructive criticisms are in order if responsibilities are not met. But neither praise nor criticism should be offered unless they are furnished in a constructive, cooperative way.

Letting Employees Know About the Security Program

It is seldom advisable for managers to inform employees about all of the details of security. At the same time, employees need to know the reasons behind company rules and requirements. It is human nature to resent discipline unless the reasons for it are understood.

For example, informing employees that employee automobiles are frequently vandalized on the company parking lot diminishes employees' resentment of careful security checks of automobiles coming on the property.

There is seldom a secret purpose behind company security rules. Employees are almost always sympathetic if they are aware of the problems that are faced by management and the company security representatives. It may be well to point out to employees that many of the company procedures and rules are directed toward protecting employees and employee property.

Reminding Employees of Security Responsibilities in Awareness Sessions

Some firms have reported that continued employee honesty can be cultivated by the use of so-called security awareness sessions. Basically, these are informal meetings between a security and/or management representative and employees to discuss security problems. As many as twenty employees can be included in an awareness session at one time.

In these meetings, a security representative or a member of management leads a frank discussion as to what the company seeks to accomplish, how company objectives may help all employees, and security rules that may be involved. These meetings can also be used for requesting security suggestions from employees.

Most employees are, of course, well aware of security responsibilities but need to be reminded from time to time. Good security usually results almost automatically when management and security representatives have created an atmosphere in which workers regard themselves as part of a company team. When this employee awareness is created, many individuals will take pride, even find stature, in devising new ways to support the company security program. If encouraged to do so, most employees will identify themselves with a business of recognized honesty and integrity.

Company Security Rules

As pointed out previously, security may usually be more easily maintained if company rules are well known to employees and are uniformly enforced. Some companies supply a list of company rules to every new employee. Others print booklets or bulletins that include all security rules, posting these on employee bulletin boards. To make certain that the rules are always available, some firms secure the firm's manual to the bulletin board with a small chain or heavy cord. Other companies stress security rules, one by one, at employee meetings. This is done either in group sessions or in small indoctrination meetings handled by the security representative.

Some firms have reduced all company rules to fit on one or two sheets of paper. Immediately after a new employee is hired, this list of rules is presented to the employee to be read and signed. Preceding the employee's signature space is a statement to the effect that the rules have been read and understood by the employee. This signed copy of the rules is then placed in the employee's personnel file. This document may be very effective in disputing an employee's subsequent claim of lack of knowledge of the company rules.

The following rules are typical of those used in many companies. It is generally made clear to employees that a violation makes an individual subject to immediate discharge or other disciplinary action. Some or all of these rules may be applicable in different businesses:

1. Discharge is the accepted penalty for theft or unauthorized removal of company property from the premises.
2. Possession or use of alcoholic beverages or dangerous drugs or nar-

Personnel Security 195

cotics on the company parking lot or inside company buildings is prohibited.
3. No employee may report to work under the influence of drugs or alcohol.
4. Destroying, deliberately damaging, or misusing company property or the personal property of a co-worker is forbidden.
5. Engaging in gambling or selling of gambling tickets, lottery chances, or other materials used for gambling is forbidden.
6. Involvement in any kind of sexual misconduct, or soliciting or inducing another to become involved, is prohibited.
7. Fighting during work hours is prohibited, as is fighting at any time on company property.
8. Falsifying production figures, payroll records, or personnel forms is prohibited.
9. Using obscene language, making threats, making malicious, vicious, or profane statements about any company official, supervisor, fellow worker, or employee is prohibited.
10. Punching the time card of another worker or inducing another employee to falsify or alter pay records is forbidden.
11. Removing or posting signs on the bulletin board without approval or leaving materials or signs on company property that reflect unfavorably on the firm or its employees is prohibited.
12. Sleeping on the job or hiding from a supervisor is prohibited.
13. Deliberately interfering with production or threatening others who will not cooperate in sabotaging production is grounds for immediate discharge.
14. Unnecessary games, shouting, horseplay, or confusion on the premises is forbidden.
15. Creating unsanitary or dangerous working conditons, or refusing to wear recommended safety equipment or clothing, is prohibited.
16. Deliberately loitering, killing time without reason, or leaving a work assignment is prohibited.
17. Deliberately smoking where such activity is forbidden because of health, safety, or security is prohibited.
18. Deliberately performing assigned duties in a way that jeopardizes the safety of other employees or visitors is prohibited.
19. All employees must enter and leave the building by the employees' entrance only.
20. All packages brought into or taken from the building are subject to inspection by company supervisors.
21. Visitors will not be allowed to come on the premises without permission.
22. Employees are required to leave the premises after checking out.

23. Unauthorized vehicles are not allowed in loading and unloading areas at any time.
24. Only company fleet vehicles are permitted to use company gas and oil or to be serviced or repaired on company time or at company expense.

DEALING WITH PROBLEM EMPLOYEES

Helping the Problem Employee

Every individual has the right to live a private, independent life. But living habits of an employee may reflect a life style beyond the earned salary of that individual. In a situation of this kind, the employee may have outside family resources, a second job, or savings; or he may be stealing from his employer. With the benefit of hindsight, managers sometimes realize a dishonest worker had been leaving clues that should have been recognized for years.

The training and replacement of employees is very expensive in almost every business. It is usually preferable to detect an employee who may be in trouble financially and emotionally and to extend counseling and help, if this can be done without interference in the person's private life.

This does not mean that a business is operating a psychological counseling service. But managers and security representatives should recognize that it may be good business to help employees with problems, at least on a selective basis.

There are always some individuals who hold up well under almost any kind of business or personal pressures. But experience shows that many workers under stress may try to solve their problems at the expense of the employer, through theft or embezzlement.

Managers frequently find that supervisors have been aware of employees' personal problems but have not brought these to the attention of management. Supervisors should be encouraged to advise management, so that a decision can be made as to whether the employee should be counseled, extended a loan, or otherwise furnished assistance.

Discharging the Thief

Experience shows that an employee who becomes involved in theft can seldom limit activities to a single incident. Once involved, an employee will almost always continue to seek new opportunities. If management forgives a dishonest employee and retains that person on the job, this information can seldom be kept from other employees. They may adopt the attitude that one employee is forgiven and allowed to remain on the job, others can expect to steal and get away with it. Most experienced security representatives agree that management should discharge a thief on discovery, regardless of the person's years of service, the personal feelings of management, or any other circumstances.

Temporary Employees

Because of absenteeism, many firms may be forced to use temporary employees. While this is often a temporary solution, it is seldom satisfactory from a security standpoint. In most instances, a temporary employee will have little loyalty to the company, which increases the likelihood of engaging in internal thefts.

If it appears that a worker has not been able to hold a steady job, there may be cause to determine why. An individual simply may not choose to work full time. However, an employee who consistently seeks temporary work may be an alcoholic, a confirmed drug addict, or just plain lazy. This does not mean that temporary employees are consistently dishonest or without ability. Some merely lack the ability to sell their skills on the labor market. At times, temporary workers who prove themselves may be hired on a permanent basis. If this is done, it is suggested that temporary workers be processed through regular company employment procedures and background investigations just as any other applicant.

It is to be emphasized that temporary workers, by the very nature of their relationship with the business, usually have limited loyalty to the employer. In most instances, they have little or nothing to lose if they should be discharged. Temporary workers seldom form close personal relationships with co-workers and are unlikely to be influenced by rank and file opinions as to honesty and reliability.

Temporary employees are sometimes hired through an outside agency; experience shows that often such agencies have insufficient background information on these workers. If a security investigation needs to be undertaken for any reason, a firm may need to contact a former temporary employee. Thus the firm should keep a record of these workers' Social Security number, birth date, state driver's license number, home address, and telephone number.

Some Examples of Problems Caused by Temporary Employees

Recently, in a western state, a temporary worker stole a number of transistor radios from a warehouse. Unsupervised, this individual was able to make repeated trips with stolen items to the outside. Concealing these radios in shrubbery on a nearby lot, the thief returned after dark to pick up the loot. The thief removed each item taken from its individual package, leaving the empty boxes on the warehouse shelf. Eventually, the thief pawned the radios for enough money to leave town. The empty merchandise boxes were not found on the warehouse shelves until an inventory was taken several weeks later.

Some firms have found that the most effective way to control nonpermanent workers is to assign a regular employee to be responsible for one or more temporary workers. Such an assignment is not practical in all situations, however, especially if the job can best be performed by a lone employee. But in unloading merchandise from a boxcar, for example, one regular employee may be able to supervise the work of two or three temporary workers. The success of this kind of supervision depends on the reliability of the regular employee and an understanding of the extra responsibility.

Recently, a temporary worker in a Miami, Florida, warehouse picked up the warehouse superintendent's key ring when no one was observing his activities. The temporary worker than slipped away to a telephone, calling an acquaintance who came to the warehouse and picked up the keys. The acquaintance then duplicated the keys and returned them without anyone's being aware that they were missing. A night or two later, the duplicate keys were used by burglars to gain access to the building.

A temporary worker in the St. Louis area noted that door padlocks were left unsnapped in the warehouse where this individual was temporarily working. Hired to work at this location a second time, the temporary worker brought a used padlock of the same brand that he had seen hanging unlocked on the warehouse door hasp. Pocketing the company lock, the temporary employee replaced it with the padlock he brought, which appeared to be identical. When the warehouse superintendent locked up for the night, he used the substituted padlock. At the close of the work shift, the temporary employee managed to remain in the building by hiding in the men's rest room. Several hours later, the temporary worker came out of hiding, opened the warehouse door with the key to the substituted padlock, and removed a large amount of merchandise.

In still another case, a temporary worker in Seattle released a bolt that secured a basement window. The following night, the unlocked window gave access to a burglar who was working in collusion with the temporary worker.

A warehouse in Phoenix, Arizona, hired two temporary laborers. The warehouse superintendent making a required monthly check of building security, inspected the railroad spur track that ran alongside the warehouse dock. He observed loose wiring that protruded under an exterior warehouse door. Looking for the origin of the loose wiring, he discovered that the temporary employees inside this warehouse had short-circuited the alarm on the warehouse door. They had also removed the lag bolts that held the hasp on the lock securing the warehouse door. Inside the warehouse were a number of valuable color TV sets.

The warehouse superintendent immediately notified police of his discovery. The police staked out the warehouse during the early night hours and arrested the two temporary employees who had come back to the warehouse for the purpose of committing a burglary.

The security representative should insist that management be selective in the kind of work assigned to nonpermanent employees. If there is a "high-value" area in the warehouse, it should be placed off limits to temporary workers. Experience also indicates that temporary employees should be provided temporary badges with photographs and identification. This will enable regular employees to quickly identify temporary workers inside the business. Temporary employees should be held accountable for the badges at the completion of the working day.

Exit Interviews

An exit interview can be a valuable tool in a personnel security program. Of course, an exit interview is not possible if the employee is discharged and leaves before an interview of this kind can be scheduled. An exit interview gives the employee an opportunity to list grievances and to report troublesome situations that should be brought to the attention of management. As a result of exit interviews, management often learns of problems it had no idea existed.

If the departing employee is hostile, little information of value can be expected. Frequently, however, a skilled interviewer is able to establish good rapport with the departing employee. In the course of their discussion, the departing employee can be asked quite frankly of knowledge of situations involving theft or other misconduct that should be known by the company. In all interviews of this kind, the employee should be told that any information provided will be kept in confidence.

An exit interview may also be helpful in reducing loss if the interviewer can go through a check list to obtain the return of confidential manuals, keys, restricted information, technical books, vehicle stickers, credit cards, or other items belonging to the company.

Then, too, if the employee has had access to trade secrets or confidential records, an exit interview can be used to remind the departing individual of responsibility to keep this information confidential. Approached properly, the departing employee will seldom violate trust. Most interviewers here simply point out that this employee has been worthy of trust in the past and that the company is convinced of the employee's basic honesty and is confident that this trust will not be misplaced in the future.

Bonding

In setting up a security program, consideration should be given not only to avoiding embezzlement, theft, and other losses, but also to recovering such losses if indeed they occur.

Restitution is frequently possible if the theft or embezzlement is committed by a company employee. In some instances, management is quite surprised to learn of employee dishonesty and can even be shocked into inactivity, but recovery should be obtained from the dishonest employee when possible.

Some firms protect themselves in advance by requiring employees handling cash or valuables to be bonded. Perhaps no more than 15 to 20 percent of all U.S. firms go to the expense or trouble of requiring bonds, however.

Background checks by a bonding company will sometime reveal character traits previously unknown. These investigations may also have the affect of discouraging future misconduct. The bonding process frequently works as a psychological deterrent, alerting the employee that management has thought about the possibility of misconduct in connection with the employee's particular job.

Bonding, however, is not intended to replace good control procedures against theft or embezzlement. Bonding is intended merely to serve as a type of insurance in case the company controls prove ineffective. Some firms, however, make the mistake of assuming that since employees are bonded, a good security program is not necessary.

A Detroit firm attempted to save money by not bonding the company bookkeeper. Management reasoned that the bookkeeper, a kindly looking, gray-haired woman, would never become involved in misconduct. Within a period of two years, however, this woman stole approximately $50,000 from the firm by altering company checks. When the theft was uncovered, the company was forced to go into bankruptcy.

A large Illinois firm required all employees handling money to fill out a bonding application. Deciding to save money, the firm merely filed the bonding applications after they were completed. Management reasoned that this would work to the benefit of the company, since employees were under the impression that they were bonded. Eventually, however, it became common knowledge among employees that the bonding fees had not been paid and that the bond application was merely a bluff. In the end, this worked to the disadvantage of management and the company security program.

Management and security representatives should realize that bonding alone does not automatically confer protection on the company. Normally, the bonding company will decline to pay for losses which cannot be verified by an outside Certified Public Accountant or otherwise documented.

At times, the cost of obtaining evidence to support a bonding claim may be more than the provable loss. A resourceful security official can sometimes obtain the cooperation of the guilty employee in locating proof and obtaining admissions.

If the loss involves a merchandise shortage, the prospects for recovery usually depend on being able to prove how the loss took place. The maintenance of good inventory records may be useful in obtaining a compromise from the bonding company in some situations.

Experience indicates that it is best to notify the bonding company as soon as a fraud or theft is suspected. It is also suggested that management or the security department consult with the bonding company before trying to obtain restitution from a dishonest employee. If a thief or embezzler is allowed to sign a promissory note for the money or merchandise taken, recovery on the bond may not be legally possible. This depends on the wording of the bond and on the facts in the particular situation.

If a promissory note is accepted from the thief or embezzler, it is not unusual for the subject to disappear after making a few payments. When this occurs, the bonding company usually is relieved of any liability to make up for the remainder of the money owed by the criminal.

It is to be noted that bonding (fidelity insurance) is usually a valuable protective aid to business. But bonding should never be considered as an alternative to an effective security program.

SUMMARY

Most physical security devices and systems may prove ineffective unless honest employees are available to use them. It is therefore essential to hire honest employees and to encourage them to remain honest. Stability and dependability are highly desirable traits in employees. Drinking, marital problems, unpaid bills, uncontrolled credit buying, habitual absenteeism, job switching, and frequent changes of address may not be conclusive proof of instability, but all can point in that direction.

Federal laws, designed to eliminate discrimination, limit the types of questions that can be asked of applicants, but a skillful interviewer can extract enough information to indicate whether the job seeker is unstable. Lack of stability may be indicated by marital problems, uncontrolled buying or unpaid bills, bankruptcy, habitual absenteeism, job switching, wage garnishments, and other personal problems. Experience shows that employees may resort to embezzlement or theft if personal pressures become too great.

Perhaps the most effective way to obtain good background information on an applicant is to send out a company investigator or to hire an outside investigative agency if company employees are not available for this purpose. Personal contact with former employers and neighbors is almost always more effective than a telephone call or a form letter.

In many parts of the United States it is unlawful to require a job applicant to take a polygraph test (commonly known as a lie detector test) to learn something about the applicant's background and suitability. If an applicant volunteers to take the test, however, and if it is administered by a skillful operator, polygraph results may enable the firm to hire the applicant almost immediately after interview. The Psychological Stress Evaluator (PSE) test is also being used by some businesses that do not want to invest the time or money involved in an extensive investigation. Both the polygraph and the psychological stress evaluator have their backers and detractors, since neither test is foolproof.

An effective security program usually seeks to motivate employees toward continued honesty by letting them know that management relies on them and by requiring high performance standards. Employee motivation is strengthened if management officials set a good example. Explaining the need for company security rules may also help. Some companies remind workers of security responsibilites in so-called security awareness sessions, in which objectives and needs of the security program are explained. Experience shows that employees are far more likely to accept security regulations if they understand the problems that management has.

Without prying into the personal lives of employees, managers and supervisors should maintain enough contact to be aware when employees get into difficulties that could lead to embezzlement or theft. Experience shows that it is usually easier to extend some help and sympathy to an employee rather than to train a replacement.

Temporary employees are usually more likely than permanent

employees to steal or embezzle company funds because they do not have the same loyalty to the company and normally have less to lose.

Exit interviews with all departing employees may pay dividends in obtaining the return of company materials, property, keys, and confidential manuals. In addition, they may be used to obtain any information departing employees can give as to employee dishonesty or security problems.

Employee bonding or fidelity insurance may be a valuable technique if used to back up the company's security program, but bonding should never be used to replace regular security measures. There are times when bonding may be illusory, because the bonding company will not pay off on a claim unless there can be a definite showing that the bonded employee was directly responsible for the loss of money or merchandise. In many cases, this direct responsibility is very hard to prove unless the guilty person makes a complete confession of wongdoing, including specific details that can be tied to the loss picture.

REVIEW AND DISCUSSION

1. What problems may be anticipated because of failure to hire honest people? How does personnel security relate to physical and procedural security? Explain.
2. List some possible indicators of instability that may be uncovered in a job applicant's background. How does this instability affect an employee from a security standpoint?
3. What are the objectives in an investigation into the background of an applicant for a job?
4. How does a polygraph work? Describe some of the benefits of and objections to the use of a polygraph in preemployment testing.
5. Explain how the Psychological Stress Evaluator (PSE) works. Does it make physical contact with the person undergoing the test?
6. What techniques can be used to motivate employees to remain honest?
7. What are employee awareness sessions? How are they conducted?
8. List ten typical security rules for a business or institution.
9. What are the security risks in hiring temporary employees?
10. Explain how exit interviews may be used to enhance the company's security program.
11. Outline some of the benefits obtained from bonding employees.
12. Does bonding of employees automatically protect a business against employee theft? Why or why not?

13

Computer Security Problems

Security, as it relates to computer installations, is a specialized field. There are a number of books that deal exclusively with this area of security. In many respects, physical protection for a computer installation requires the same kind of planning businesses use to prevent burglary, theft, or other crimes against property. But protection for a computer facility may require some measures that are not common in other security programs. There is no question that extensive technical knowledge of computers is helpful in devising computer security plans. In most instances, however, a competent security employee in combination with computer professionals can devise a well-rounded program. The purpose of this chapter is to outline protective features that can be used in security programs of this kind.

BACKGROUND

During the 1960s and early 1970s, many business firms and institutions replaced existing record-keeping systems with computer tapes or disks. As businesses and firms have made additional use of the computer, they have become more dependent upon electronic data processing records. Experience has proven that loss of these computer records can destroy an entire business. Accordingly, the vulnerability of computer records cannot be overemphasized.

The most effective programs are those that are based on firm rules written out as clearly and concisely as possible. In addition, regular reviews and inspections of procedures by management and the director of loss prevention or security contribute greatly to a program's effectiveness, insuring that security hazards do not develop in day-to-day operations.

LOCATION OF THE COMPUTER INSTALLATION

In planning to avoid security problems, consideration should be given to locating the computer facility in a relatively crime-free neighborhood, where employees may report to work at all hours without hindrance. In addition, the choice of site should be influenced by the location of neighborhood emergency support facilities, such as the fire department and the police station. No one can predict that a particular neighborhood will always be free from riots, but through study and planning, a firm can at least minimize the likelihood of choosing a riot-proof location.

Physical Catastrophes

A number of case histories document the destruction of computer systems by physical disasters. Many of these catastrophes could have been avoided through proper planning. Man-made disasters usually involving riots or industrial sabotage, have been responsible for some of these losses. Natural disasters have also played their part in the destruction of computer installations, and these have included damage from fire, flooding, building collapse, explosion, tornado, hurricane, and earthquake. Of the natural catastrophes, fire is probably the most serious threat, and it will be considered in this chapter in detail.

Against the advice of the firm's security director, a New Orleans company located the company's computer in the basement of a new building in a low-lying suburb. When a nearby creek flooded, serious damage was done to the computer hardware and many company records were lost.

A major airline recently sustained damage to an IBM computer when a pipe in the air-conditioning system used in the computer room was accidentally ruptured. The chief engineer of the building and the company director of security had both recommended that the air-conditioning units be placed in another location, but this would have involved additional cost at the time the building was constructed.

The computer room of a linen supply company in the Los Angeles area was damaged because no drainage system had been built into the underfloor of the computer installation. Electrical connections were shorted out when water poured into the computer room from a blocked drain in a nearby office.

If there is a possibility that a water main or sewage connection could flood into the data processing area, consideration should be given to a poured concrete roof or other protective features to keep the water out.

A computer installation in the Ft. Worth, Texas, area was a complete loss in 1977 when the roof collapsed. Over a period of time, roof drains had become plugged with debris. Roof areas on higher sections of the building drained on the flat roof above the computer room. When a flash flood occurred, the roof was not able to stand the weight of accumulated water.

The 1966 explosion in the Phelps Dodge Copper Company building in Ft. Wayne, Indiana, caused by leaking natural gas, was reported to be the first explosion that destroyed an entire computer system. Although the Honeywell computer was ruined, 150 computer tapes stored in the firm's new computer safe were all usable. Other explosions have occurred in computer installations in more recent years.

Although where and when hurricanes, tornados, cyclones, and earthquakes will take place cannot be accurately predicted, it can be anticipated that a natural disaster of any kind may cause serious loss. If the company has an option, it should locate its computer system away from potential trouble spots. Hurricane Carla completely destroyed a Burrough's computer at the plant of a major chemical company at Freeport, Texas, in 1963, and hurricane Celia damaged a number of computers at Corpus Christi, Texas, in 1970. A major computer center suffered considerable damage in an earthquake in the Los Angeles area in 1969. The center was located near an active earthquake fault.

Materials are sometimes brought from outlying cities for processing on a daily basis. If a helicopter is used to transport them, it is recommended that the landing space for the aircraft be on a protected, reinforced concrete roof. Even with this precaution, there is a possibility that a rough landing could spray large quantities of burning gasoline down the walls and into the building.

Costly damage to a computer occurred in a Missouri city when city officials sandblasted the stonework on the exterior of the city hall. Located inside this building, the computer was covered with minute, almost invisible particles from the sandblasting operation. As a result, the computer hardware was unusable and some computer disk packs could not be copied. If the computer and disk packs had been covered with plastic sheets, the loss probably would not have occurred.

Physical Controls

From a security standpoint, it is best to locate a computer operation in a separate building. As a practical matter, however, many businesses cannot afford a separate building. This means that security problems in the existing building must be corrected or minimized. If possible, the electronic data processing installation should be located out of the flow of employee and visitor traffic, behind substantial walls, and away from public areas. Obviously, physical controls must be tailored to the individual location.

Not only should protection be given to the building in which the computer is located, but additional security should be provided for the computer room and areas where computer hardware and records are maintained. Any device or technique that provides good security in normal business installations may be considered here.

FIRE AS A THREAT TO THE COMPUTER

For a time, business and security officials did not realize that there were critical differences between the security needs of old-style paper records and those of records maintained on computer tapes and disks. Paper documents usually remain readable after exposure to temperatures of up to 350° Fahrenheit. Above that temperature, paper documents become blackened and charred. Present-day fire safes for the protection of paper records generally make use of insulating materials that are placed between the outer and inner steel walls of the safe. When exposed to fire, these insulating materials or chemicals are changed in composition, giving off steam that protects documents inside the safe without causing serious water damage.

While polyester-backed computer tape will not ignite until a temperature of approximately 1000° Fahrenheit is reached, heat and humidity may cause damage to tapes or disks at temperatures considerably below that point. Exposure for even a short time to temperatures no higher than 150° Fahrenheit may cause computer tapes to become unreadable. Water may be sprayed over tapes at low room temperature, but humidity may cause serious damage any time room temperature reaches 150° Fahrenheit.

Attempts to protect computer records in old-style safes failed because steam in insulating materials in the safe caused the tapes to deteriorate. Computer tapes or disks must be stored in specially built computer safes to insure adequate protection.

There are a number of reasons why computer centers are easily damaged by fire. In the first place, computer systems require large amounts of electrical power for normal operation, which means a large number of electrical wires and connections that are usually located in an underfloor area. A short-circuit in any of these electrical connections or wires can cause fire. Experience also shows that large quantities of combustibles including paper forms are frequently stored in or near computer rooms.

The so-called Pentagon fire of July 2, 1959 was one of the most dramatic examples of serious damage by fire in a computer room. Three IBM computers and an estimated 5000 to 7000 reels of computer tapes were destroyed when a fire swept through the computer room and tape library in the Pentagon. The loss of confidential files was considerable. One security authority estimated that the replacement cost for these materials was $6.69 million.

Typically, a fire in an electronic data processing center may break out in areas that cannot be readily observed. These are fires under the raised flooring, in equipment cabinets, or in the area under drop ceilings. When a short-circuit occurs in one of these areas, the fire may smolder for a prolonged period. Gases from a smoldering fire of this kind may eventually choke computer room employees and obstruct visibility. Unless approved fire-resistant electrical wiring and connectors are used, the plastic and other cable coverings may give off a toxic black smoke while burning. In the confined area of a computer room, this may add to the difficulties of firefighters in reaching the blaze.

To control fires of this nature, three basic procedures should be followed:

1. Prevent fires.
2. Promptly detect and extinguish fires when and if they occur.
3. Prevent the spread of any existing fires.

To prevent the outbreak of fire, it is usually helpful to make certain that there are few, if any, combustibles in the computer room. Underfloor cables should be of the type approved by Underwriters' Laboratory. These cables are covered with flame-retardant materials and so-called self-extinguishing insulation and jackets.

A survey should be made of all rooms in the computer area. Wooden desks, flammable drapes, and flammable floor coverings should be removed or treated with sprays to reduce flammability. Wooden file drawers should be replaced with the metal type and all fixtures and decorations should be noncombustible.

Location and Use of Detectors

For good results, fire detectors should be installed at the inside of each computer or console cabinet. In addition, smoke or fire detectors should be located in areas under the floor, as well as in ceiling areas, in exhaust and air return ducts, and in other locations where fire is likely to break out. In a large computer installation, a grid control system is sometimes installed, showing the location of the specific fire detector that was activated and sounding an audible alarm. By observing the grid, the guard or computer employee at the control box can immediately determine the location of the fire even though it may be in the underfloor area.

Ionization type detectors are usually recommended as preferable for computer rooms. These detectors are activated by changes in the gaseous content of the air or by variations in air currents. Ionization detectors are very sensitive to chemical particles in minute quantities of smoke or dust and may be set off accidentally by some types of cleaning solvents. The smoke from a single cigarette is usually sufficient to activate a detector of this type.

Ideally, the computer should be equipped with an automatic shut-off device that will alert the computer room operator and shut off electrical power if an ionization detector is activated. Experience shows that if the electric power is turned off at the time of a fire, the computer is considerably less likely to be damaged. In addition, employees are less likely to be electrocuted if the power is off when water is sprayed on open electrical connections in the computer room.

It should also be pointed out that most computer installations rely on air conditioning systems to maintain a low temperature in the computer room. Without these air conditioning systems, many computers would become overheated and inoperable. At the same time, there is a possibility that an air conditioning system may blow fire or smoke throughout large areas of the building in a short time. It is therefore important for the air conditioning units to be shut off automatically in the event fire is detected or for the air conditioning ducts to be closed off with a damper.

Extinguishing Agents for Computer Fires

Three types of fixed fire-extinguishing systems are regularly used to put out fires in computer areas. Opinions differ widely as to which is superior. These three types are:

1. Automatic water sprinkler systems.
2. Carbon dioxide (CO_2) gaseous system.
3. Halon gaseous systems (usually Halon 1301, known by the duPont Chemical Company trade name Freon.

In addition, high-expansion foam and dry chemical extinguishants have been used to put out computer fires. Foam and dry chemical extinguishants have not been widely used, however.

Water Sprinkler Systems

Automatic water sprinkler systems have been widely used to extinguish business and industrial fires of all types. This extinguishing system makes use of water-filled pipes throughout the ceiling of the building. These pipes are equipped with sprinkler heads that open up and spray water on the area below the sprinkler head when a preset temperature (generally about 160° to 180° Fahrenheit) is reached. The system is activated by the melting of fusible links in the overhead sprinklers. One of the weaknesses of this type of installation is that air conditioning currents may dissipate the heat generated by a fire. As a result, floor-level temperatures may be extremely high before the fusible links in the ceiling pipes are hot enough to melt and open up the ceiling sprays. Once started, the spray of water from an automatic water sprinkler system will continue indefinitely until the water supply is cut off.

There are a number of advantages to a water sprinkler installation in fighting a computer fire, but there are also serious objections. Compared to other systems, the water sprinkler system costs little to install and maintain. In the event of a serious fire, the stream of water continues to spray as long as necessary. In addition, water cools down the burning materials so there is little possibility of a "flash back" or a second outbreak of flame as soon as use of the extinguishing agent ceases. Water sprinklers are also usually reliable and trouble-free in maintenance and operation.

But a spray from an overhead sprinkler is not likely to reach a fire inside a computer cabinet. And even more important, steam is almost always the result when a spray of water strikes a fire. As previously pointed out, steam, even at comparatively low temperatures, will consistently ruin computer tapes or disks. Another problem with sprinkler systems is that water causes short-circuits in electrical connections and the computer and other pieces of equipment may be seriously damaged if electrical power has not been disconnected. There is also a possibility that an employee may be electrocuted if the spray from a sprinkler head strikes an active electric circuit in the computer room.

Water also will cause electronic components in the computer to rust within a very short time. If water is used to put out a computer room fire, it is very important that the computer hardware and other pieces of equipment be dryed out as quickly as possible. If this cannot be done with large blowers or fans, serious damage to the computer can be anticipated in a matter of a very short time.

Carbon Dioxide Gas (CO_2)

Experts on fire prevention have long used carbon dioxide gaseous systems in some industrial installations. Carbon dioxide is not a conductor of electricity, and there is little danger of anyone's being electrocuted if it is used.

In addition, carbon dioxide does not corrode computer terminals nor ruin sensitive computer equipment. If a fire is extinguished with carbon dioxide, the computer equipment can usually be put back in operation just as soon as the fire is out and the carbon dioxide has been blown away.

It must be pointed out, however, that carbon dioxide is a deadly gas in the concentrations that are used to put out a computer room fire. If an employee should become trapped in the computer room or fail to leave in time, it can be expected that the employee would be asphyxiated. Consequently, carbon dioxide should not be installed without adequate training programs for computer room personnel as to the dangers involved. In addition, one or more supervisors should be trained in the use of a self-contained breathing apparatus that could be used for rescue.

In installations where carbon dioxide is used, it is necessary to sound a loud alarm prior to discharge of the gas, to allow employees time to leave the computer room. This delay, of course, allows a few additional seconds for the spread of the fire.

One of the basic weaknesses of a carbon dioxide system is that the spray of gas will not continue indefinitely. The carbon dioxide is stored in cans or bottles, and the gas gradually dissipates after the supply of carbon dioxide has been exhausted. In effect, carbon dioxide smothers the fire by cutting off the oxygen it needs to keep burning. Unlike water, however, the carbon dioxide does not cool down the materials that are burning, and there is always a possibility that a fire will again flare up after the carbon dioxide has dissipated.

The installation of a carbon dioxide gaseous discharge system is more expensive than that of a water sprinkler system. Another potential drawback is that carbon dioxide creates a cloud in the atmosphere, obstructing visibility in the area where the gas is discharged. This may hinder fire-fighting efforts.

Halon 1301 (Freon)

Halon 1301 is usually known by the trade name Freon, under which it is sold by the duPont Chemical Company, or simply as Halon. There are a number of extinguishing agents in the Halon chemical group, and the number 1301 is given to the type that is generally used to extinguish computer fires.

Halon is a colorless, odorless gas and consequently does not obstruct the visibility of those trying to get at the fire. In addition, Halon eliminates the possibility of electrocution, since it is a nonconductor of electricity.

Chemists are not certain how Halon works in putting out a fire. There has been no reported instance of death from exposure to Halon in a computer room fire, and it is therefore considered far safer than carbon dioxide. There are a number of cases on file where computer employees remained in the computer room following a discharge of Halon sufficient to put out the fire.

The National Fire Protection Association (NFPA) in Boston, Massachusetts, has long set the standards in this country for fire prevention. For a time, officials of NFPA were somewhat reluctant to approve the use of Halon, although it had been strongly recommended by most computer manufacturers. Tests that have been conducted reflect that Halon may chemically decompose at the high temperatures encountered in a fire, and that the resulting products of combustion could possibly be dangerous to humans. Halon has been used in computer fire after computer fire, however, since the NFPA first expressed reluctance to approve Halon. No death or harm of any kind has been reported from these actual applications of Halon in computer room fires. As a consequence, computer manufacturers almost uniformly prefer Halon to water sprinkler systems or carbon dioxide.

It should be kept in mind that if the fire is of the type that smolders, then neither Halon nor carbon dioxide may be adequate to prevent a flashback after the available gas supply has been discharged. For this reason, some computer room installations use a second-shot system, a backup supply of carbon dioxide or Halon that can be introduced into the computer room if necessary.

Halon is quite expensive, usually costing considerably more to replace than carbon dioxide.

As the 1975 National Fire Protection Association Bulletin #75 points out, NFPA now recommends water sprinkler systems, carbon dioxide, or Halon 1301. Some fire insurance companies, slow to change, still require the use of water sprinkler systems in computer installations. As noted, this is almost always against the advice offered by computer manufacturers, who can cite numerous cases where computer hardware has been ruined by water sprinklers.

Hand Extinguishers and Floor Pullers

Regardless of the type of extinguishing system used or recommended in the computer room, authorities on fire prevention uniformly recommend that hand extinguishers be available throughout the entire computer installation. Most authorities recommend carbon dioxide or Halon 1301 extinguishers. In the quantity available in hand extinguishers, carbon dioxide is not likely to prove dangerous to human life.

The computer installation should regularly train all employees in the use of hand extinguishers and other protective devices against fire.

Experience shows that hand extinguishers should be mounted on mounting brackets or hooks where they are readily available and in locations where the absence of an extinguisher would be immediately noted. Some experts on computer fire problems recommend that a pattern of a distinctive color (usually red) be painted on the wall behind individual fire extinguisher brackets. They also suggest that the director of security or loss prevention institute a program to regularly check the serviceability of all hand extinguishers, especially those in the computer area.

Floor pullers (with a handle in the middle and rubber suction cups at each end) are commonly used to raise up the floor tiles in computer rooms. Without these pullers, it is usually very difficult to gain access to the underfloor area in a computer room, where most electrical cables and electrical fires are found. It is recommended that floor pullers be mounted on brackets alongside hand extinguishers. If floor pullers are not available, it may be very hard to raise a tile and direct a hand extinguisher to the location of a fire.

Floor pullers can be used to make periodic inspections of the underfloor area. This area should be regularly vacuumed or cleaned. If it is not, candy wrappers, sweepings, and trash may accumulate, presenting another fire hazard.

BACKUP RECORDS

Off-Site Storage of Backup Records

Experience shows that all vital computer records are usually destroyed in any disaster that is serious enough to wreck a computer. It is usually possible to obtain a new computer, but records can seldom be replaced. Statistics maintained by insurance companies for many years reflect that most businesses are not able to continue to function if vital records are lost in a disaster. Ideally, exact duplication of computer records is desirable. From the practical standpoint, exact duplication is usually too costly for most businesses and institutions that maintain a large number of records.

Most computer installations solve this problem by using a so-called three-generation backup system, storing the backup materials at a separate location. A scheme of this kind is sometimes known as the "son-father-grandfather" method. The "son," or up-to-date file, is a combination of the "father," or previous file, plus those transactions necessary to make the file current. The computer run used prior to the "father" is designated the "grandfather" and is usually kept at an off-site storage location. On a regular basis, the "great-grandfather" file is destroyed, leaving the three generations intact. By keeping the "grandfather" file at a separate location, the company always has a backup that can be obtained in the event of fire or other disaster.

Most authorities on computer protection feel that the off-site storage location need not necessarily be a long distance away from the computer installation. The backup site should, however, be sufficiently far away so that it would not be affected by a fire or other disaster at the computer installation.

As a precaution, the company should assign responsibility for the backup tapes to an independent employee who has no access to the primary tapes or disks at the computer center. This is so that no one individual can destroy both the primary and the backup tapes and materials.

Backup Systems

While computers and other units of hardware in the computer installation are usually reliable, they are subject to mechanical failures. Most businesses cannot remain in operation for very long without an available computer system. Therefore, it is usually advisable to set up backup systems that can be used in the event of an emergency.

Backup systems are usually needed for the computer (computer processor and all peripheral machines or hardware), for air conditioning units, and for an alternate source of electrical power. Through trial and error, it has been found that dry runs of these systems are advisable. Frequently the computer hardware relied on for a backup system may not be compatible when actually put to the test. Air conditioning backup is usually necessary to make certain that required temperature conditions can be maintained in the computer room. If the need to keep the computer system operating is unusually critical, it may be desirable to install a backup generator, with an adequate fuel supply (diesel oil or other industrial fuel) protected so that the backup generator can be used if the usual source of power is lost or overburdened.

FURTHER PREVENTIVES

Access to the Computer

Experience shows that physical access to the computer installation should be limited to only those persons who have a need for access because they are involved in activities that support the computer operation. Access to the computer room proper should be further restricted. Computer programmers, for example, should not be allowed in the computer room unless they are working under supervision. A programmer can change existing computer tapes to suit himself or herself and then later program out the changes, leaving no record of tampering with company records.

Tape Library Controls

An effective security program should make use of controls over the issuance and use of computer tapes or disks, with logs or other controls maintained and regularly audited. Accountability should be set up so that a theft or duplication of documents or programs would be immediately noted.

Disposal of Computer Printout Materials

It is also important to set up a system to prevent computer printouts from becoming available to outside persons. Computer tapes or disks that are no longer used should be appropriately destroyed. A paper shredder may be used to destroy computer printout material that is confidential. Precautions should also be taken to make certain that the carbon paper from continuous print outforms or carbon ribbons is not available to unauthorized individuals.

Emergency Shutdown and Recovery Procedures

Actual experiences have demonstrated the desirability of setting up an emergency shutdown and recovery program. To be effective, a program of this type should be planned and coordinated by the official in charge of the computer installation, a representative of management, and the company security representative. Experience shows that a dry run of the emergency shutdown program should be made and regularly reviewed, to make certain that protective devices are still available and that employees involved in the program are aware of their responsibilities.

SUMMARY

Computer security poses some special problems. But in many respects, physical protection for a computer installation follows the kind of planning used to protect other business property from intruders or criminals.

Ideally, the computer location should be near police and fire protection facilities, in a relatively crime-free neighborhood. Women employees are always concerned with the safety of the neighborhood if they work the night shifts. Security guard patrols may be needed for parking lot duty.

The likelihood of serious damage from flooding or other natural castastrophes should be considered when choosing the computer location. In addition, most authorities on computer security feel that the computer should be housed in a separate building, where access is easier to control. This often represents an expense that cannot be justified, so an area inside an existing company building should be chosen on the basis of how easily access controls can be set up.

Fire is the most serious threat to the entire computer installation and records. Computer tapes or disks will usually be ruined at a temperature of 150° F, if steam results from using water to fight fire. Heat, steam, and smoke can all be expected to ruin the computer proper.

Ionization type fire detectors are recommended by the National Fire Protection Association for computer installations; these detectors should be located in underfloor areas, in the ceilings, inside air conditioning ducts, and in computer hardware cabinets.

In addition to hand extinguishers, three types of automatic discharge systems are usually used to put out computer fires:

1. Overhead water sprinkler systems.
2. Carbon dioxide (CO_2) discharge gaseous systems.
3. Halon 1301 (Freon) discharge gaseous systems.

There are advantages as well as drawbacks to all three of these fire suppressant systems. The water sprinkler system is the most likely to put out any kind of fire, including the kind that smolders. The sprinkler system is easy to maintain, usually foolproof, and relatively inexpensive. But steam from the water discharge may ruin tapes or disks and will quickly corrode sensitive parts inside the computer. Unless the computer can be blown dry immediately after a fire, the computer hardware may be lost. Water is also a conductor of electricity and an employee may be electrocuted if water hits an active electrical connection in the underfloor.

Carbon dioxide (CO_2) will not damage the computer or tapes and disks and does not conduct electricity, but such a system is more expensive than a sprinkler system, harder to maintain, and may asphyxiate anyone left in the computer room after the discharge begins. Carbon dioxide puts out the fire by smothering, and a deep-seated fire may flare up again after the gas is dissipated.

Halon 1301 (Freon) is far more expensive than other systems and is more difficult to maintain than water. It does not conduct electricity and probably will not damage tapes, disks, or computers. Humans have not been seriously harmed by Halon, but it may not put out a deep-seated fire that smolders for a long time. Numerous tests have been made, and Halon 1301 has been accepted by authorities on fire protection. Water sprinkler systems are still recommended as preferable by some major fire insurance companies, but there is general agreement among computer manufacturers that a Halon 1301 system is better.

Floor pullers should be available along with hand extinguishers for promptly reaching and combating fires. Employees should be trained in the use of hand extinguishers.

Most authorities on computer security believe that computer records should be backed up by prior versions of tapes or disks, along with the programs needed to operate the computer. Generally a third-generation tape is used, stored at an off-site location that would not likely be harmed by a catastrophe at the primary computer location. Ideally, the backup tapes or disks should not be available to employees at the computer center.

Backup facilities for electricity, air conditioning, and computer hardware are desirable, especially if continued operation of the computer is critical.

Access to the computer room and the tape library should be closely controlled to maintain good security. Sensitive computer printout material should be destroyed with an adequate paper shredder after the materials are no longer needed.

REVIEW AND DISCUSSION

1. Is there reason to feel that a company using computer records is more vulnerable to serious loss than a business with conventional record systems? Give reasons for your answer.
2. In setting up protection, why is the location of the computer facility important?
3. Explain some of the disasters that have destroyed or damaged computer systems.
4. Discuss the problem of fire as a threat to the computer.
5. Explain how fire detectors should be located and used in the computer area.
6. List advantages and disadvantages in using an automatic water sprinkler system to put out a computer fire.
7. What are the arguments for and against using a carbon dioxide (CO_2) extinguishing system in putting out computer fires.
8. Discuss the advantages and disadvantages in using Halon 1301 (Freon) as the agent for putting out computer fires.
9. Explain in detail why a computer security program should consider use of an off-site storage location for backup records.
10. Outline which information, equipment, or systems should be backed up in establishing a disaster recovery plan for a computer installation.

14

Protecting Trade Secrets and Confidential Information

The purpose of this chapter is to describe business espionage and how it has been conducted in the business world. Industrial and commercial procedures are often unique and may be of great value if revealed to a competing firm. Thus information concerning a product, and techniques for preparing and marketing it, may require as much protection as more visible or tangible assets such as money and merchandise. This chapter outlines how protective plans may be set up and supervised by the security department and management officials to reduce the likelihood of business or industrial espionage.

Finally this material considers the need for emergency security planning and the problems posed by bomb threats and disasters.

THE SECURITY THREAT OF BUSINESS ESPIONAGE

The Nature of Industrial Espionage

Every business organization has information that could be of real value if it fell into the hands of competitors. New product information, performance and cost figures, secret formulas, cost reduction methods, and production techniques are only a few of the matters of interest.

Some of this desired information may be obtained through business and marketing channels, but some may be obtained only through business espionage. And often, espionage poses problems that cannot be solved by installing a better lock on the door or by hiring an additional guard.

Businessmen sometimes go so far as to claim that business or industrial espionage simply does not exist. At the other extreme, newspaper accounts seem to imply that practically all businesses spend great amounts of time and money in spying on each other. The extent of business espionage and for whom it is conducted are obviously trade secrets in themselves but the existence of spying cannot be denied. Cases of proven espionage are regularly exposed in the courts. Most firms may be willing to do without illegally obtained information, but this does not mean that businesses do not need to protect themselves.

What Is Business Espionage?

Business espionage involves the unauthorized transmission or theft of closely guarded business information, or the taking of a trade secret.

A trade secret, in turn, is defined as a secret process or formula, limited only to certain individuals using it in preparing some item of trade having commercial value. In other words, a trade secret may be a device, formula, or method for treating ingredients in the preparation of a salable product. A trade secret is not necessarily complicated or highly technical. At times it may involve nothing more than a series of steps in combining ingredients, or it may make use of secret ingredients that are not commonly known to competitors. It can be quite simple and easily duplicated, so long as it is secret and produces an unusual result.[1]

The Protection Philosphy of the Courts

Businesses frequently utilize trade secrets in manufacturing or producing their product. A new mechanical device or invention can generally be protected by a patent. But information, alone, is not subject to protection under U.S. patent laws.

A patent provides protection for a mechanical device, since the device cannot be duplicated without violating the patent laws. A trade secret, however, is something that the holder must keep secret or lose the benefit to competitors.

The British and American courts have always believed in protecting

---

[1] Victor Chemical Works v. Iliff, 299 Ill. 532; N.E. 806. See also Progress Laundry Company v. Hamilton, 208 Ky. 348, 270 S.W. 834.

research and trade methods, as well as mechanical devices. Otherwise, neither businesses nor individuals would be willing to put their time and effort into schemes to develop new products.

There is nothing to keep a competitor from buying the finished product to analyze this product and to make all manner of tests on it. But a competitor cannot legally obtain the secret from an employee who has been given the responsibility of protecting confidential information. If the competitor does convince an employee to reveal the trade secret, the courts will issue an injunction, prohibiting the thief from using the stolen secret. In addition to an injunction, the courts will also grant money damages to the business or individual from whom a trade secret has been stolen.

When one of these matters comes to trial, the defense usually claims that the secret had become known to a number of individuals and was therefore no longer a secret. A claim of this kind is not enforced by the courts. Even if a company finds it necessary to disclose the trade secret to a number of employees who work with the material or processes, the large-scale disclosure does not automatically destroy the confidential nature of the secret. The test is whether management tries to keep the secret confidential within the company. Probably the best known example in American business where a secret has been successfully guarded for many years is the formula of Coca-Cola.

There are limitations on this protection philosophy, however. A business cannot claim everything done throughout the whole company as a trade secret. A specific device, method, or formula must be singled out, and steps taken to see that it is treated as secret.

The courts will also protect the secrecy of so-called negative information. It sometimes appears to experts in a particular industry or business that research into a specific area seems promising but, later, it turns out that this effort is not practical and research should be discontinued. This negative information (research that did not pay off) may be of great value to a competitor by saving large expenditures for useless research and development.

Business Espionage Is Not New

Businesses have attempted to gather trade secret information from competitors for many years, as well as to obtain secret patentable information.

To illustrate what has happened in the past, in 1780 a drunken pattern maker in the hire of celebrated Scottish Inventor James Watt bragged that Watt had discovered a way to obtain circular motion from a reciprocating piston. Other engineers and machinists who overheard this conversation in a pub expressed disbelief that this was possible, and the pattern maker chalked a sketch of Watt's mechanical contrivance on the bar. One of the bystanders was James Pickard, a Birmingham button manufacturer who realized the mechanical possibilities of Watt's contrivance. Wasting no time thereafter, Pickard went to the patent office in London and obtained a patent for this device as the "crank and connecting rod." Although there was little question that Watt had been the inventor, the courts upheld the patent right of Pickard.

Two Basic Security Problems

Generally, there are two basic security problems that face business:

1. Preventing theft of sensitive information.
2. Preventing employees from taking secrets when they switch to other jobs.

What Is Usually Taken in Business Espionage

Information that is frequently stolen from a business pertains to new product research, general know-how, pricing activities, research techniques, sales strategy, formulas, manufacturing processes, product specifications and engineering data, salesmen's compensation, plant capacity, bid information, and mergers and contract talks.

Legitimate Sources of Competitive Information

Many businesses cooperate in an exchange of information that works to the advantage of all firms in a particular industry. Generally, this cooperation results in higher product standards, an increased rate of technological development, and public confidence in the industry as a whole. But since most companies are in business to make money, they attempt to protect unusual techniques or capabilities that give them an advantage over other businesses.

There are many sources that may be used to obtain legitimate competitive information.

1. If a catalog or specific test data is published, this may be requested by anyone.
2. It is a common practice in some industries for computer managers or engineers to exchange certain types of information.

3. Considerable information may be gathered from legislative hearings, news releases, court records, annual company reports, government publications, and technical trade journals. Engineering and technical societies frequently publish information that could be regarded as confidential.
4. Comparison shopping is a technique frequently used by one retailer to obtain information about another's products.
5. Then too, valuable information can sometimes be obtained by reverse engineering or analysis of the competitor's product that is bought openly on the market. Customers will also sometimes report information on a product's performance.
6. Some businesses obtain such information by attending trade shows or open house exhibits offered by competitors, or by interviewing employees of a competitor at a professional or trade meetings.

How Secret Information Is Developed

Undoubtedly there are times when "cloak and dagger" techniques are used by business spies. However, in perhaps the majority of cases, information may be developed through relatively simple techniques if business officials refuse to take common-sense precautions.

For example, it is a relatively simple matter to hire a maintenance employee who will deliver the contents of an executive's waste basket, possibly containing items of considerable interest, to a competing firm.

Talkative salesmen and marketing employees may frequently disclose valuable information, without realizing it. At times, an overly zealous marketing representative may tell too much about a product in order to secure a sale. Even purchasing orders may contain very sensitive information, if not a total picture of the progress and research in progress at a given installation.

Consultants who move from company to company can also reveal information that should never be disclosed. Some firms attempt to restrict this type of information by requesting consultants to sign an agreement to reveal no information of any kind to competitors. An agreement of this kind may be not only legally binding on the consultant but also serve as a psychological deterrent.

Some firms have lost information so consistently at trade shows and conventions that company representatives are cautioned before attending meetings of this kind. Scientists are especially inclined to disclose information to their counterparts in other businesses, as proof of their scientific productivity and achievement.

Experience shows that a supplier can often make a significant contribution to a new development if aware of the project being pursued.

224 Protecting Trade Secrets and Confidential Information

Providing the supplier with too much information can be dangerous, however. The prudent course is to tell the supplier what that individual needs to know, with no more details.

In-house publications are frequently written with an internal audience in mind. A certain number of outsiders always read these publications, however, and sensitive information should not be reported in company magazines.

Disloyal employees are among the best available sources of secret company information. If a moonlighter works for a competing firm, the matter is one that should be of real concern to the primary employer.

Document Protection

Sensitive company information is often set out in memos, reports, or statistical summaries. Some firms have found it helpful to review these documents regularly and to give an in-company security classification to those of a sensitive nature, or those that contain trade secrets. Some companies make use of the standard classifications used by the U.S. armed services—"confidential," "secret," or "top secret." It is to be noted that if the firm is one that has contracts with the armed forces, the terms "confidential," "secret," and "top secret" should not be used except in accordance with strict military requirements.

Employees should be given these in-company classified documents on a need basis only. Charts, drawings, performance curves, and sales plans are among the types of documents that should be considered for classification. It is suggested that rubber stamps be used to mark classified documents "sensitive," "secret," "confidential," or with some other designation.[2]

Some firms have used typing paper with a colored border to indicate that the document is of a sensitive nature. Other companies have merely stamped the document with a rubber stamp notation, such as: "This document contains sensitive material belonging to———company, and may not be duplicated or used without specific permission."

If a document contains sensitive calculations, figures, or percentages, some companies have the document typed by the stenographic pool with a blank space left for the specific figures, which are inserted later by hand.

If documents are of great importance, some organizations designate an employee as a part-time or full-time document librarian. This librarian has the responsibility of maintaining charge-out slips for sensitive

---

[2]Throughout the remainder of this chapter, the terms "confidential," "secret," and "top secret" are not used in accordance with military usage.

documents and of making certain that the employee obtaining the document has proper authority to receive and use the material.

Another effective technique is to "compartmentalize" confidential information as much as possible. For example, the combination to the main money vault of a bank may be divided into two or three parts, with individual officers controlling only a part of the combination. The formula for an unusual product can also be compartmentalized in this manner.

Protecting Documents That Are Being Used

Safes or vaults can, of course, be used to protect stored documents. But work in progress is often the most valuable data to an industrial spy. The business thief usually wants to know about current matters, not those in the files.

Documents are usually safe enough while on employees' desks. But executives seldom have time to serve as paperwork messengers. Some companies place confidential documents in sealed envelopes, but the whole package could still get lost. A pneumatic tube system to transmit documents is usually more secure than the use of a messenger or routing-slip system.

One security problem sometimes encountered is that executives do not maintain a "clean desk" policy. Good security may involve a desk inspection procedure with reports of all classified company materials that are not locked up at the close of business. Another common problem results from the fact that rank and file employees may have access to confidential documents which are pigeonholed in executive mail slots in the company mailroom.

A professional association in Chicago experienced difficulty in controlling classified information until confidential items were typed on memo paper bordered in a distinctive red print. Strict rules were then instituted, limiting these items to executives in a closely restricted area on the second floor of the building. If a document was found on a desk outside the designated area, the executive head of the association was immediately notified. Executives stated that this system contributed substantially to an improvement in document control.

Telephone Taps

Newspaper stories sometimes imply that telephone taps are common in business espionage. No one really knows how often hidden microphones or secret telephone installations are placed in the offices of rival companies. Electronic experts are available through most major security companies in larger cities. These technical experts can make the necessary checks to determine whether microphones have been planted in executive offices or board rooms or whether a telephone line has been tapped. Some firms take this precaution, especially in the company board room before annual meetings or special planning sessions. Many individuals commonly feel that their telephone line has been tapped, usually because of unexplained noises on the line. There is no way to be certain without the assistance of an electronics expert, but it should be pointed out that telephone taps are usually not responsible for unusual static or noises on the line.

A number of commercial devices are available to scramble telephone calls in the event this should be deemed advisable.

Some Additional Techniques to Prevent Business Espionage

Many basic security techniques can be used in guarding against business leaks. The regular use of locks on office doors, file cabinets, and desks is always helpful. So too are access controls to the building and to confidential locations in the building. Prohibiting unauthorized persons from entering company buildings at night and utilizing the guards may also be helpful. In addition, top management should authorize the security department to make regular checks of the security of secret formulas and documents that have been designated for retention in a safe or vault.

A good security program should also be concerned with duplicating procedures and destruction techniques for confidential documents. There is little point in designating a document as confidential if employees are allowed to copy it at will. At one time, most businesses destroyed confidential items in a company incinerator. With restrictions against burning, companies now use some type of paper-shredding equipment. This procedure is effective; however, it should be pointed out that small shredders are not designed to handle large quantities of documents.

As a practical matter, a Camden, New Jersey, firm recently discovered that shredding did not necessarily mean illegibility of documents. The teeth on the company shredder were found to be set so wide apart that the strips emerging from the paper shredder were large enough to be pieced together.

The Conflict Between Company Interest and Employee Information

The courts are in agreement that secret company information belongs to the employer. An employee who has worked with secret materials can never erase this residue of confidential information by switching from one job to another. It is obvious that the interests of the first employer and the interests of the employee are in conflict, since the employee's ability includes experience gained in working with secret materials or processes. A former employer cannot curb the employee's abilities or mental processes, but if the employee reveals trade secrets to a new employer, both may be liable for damages.

Management Programs to Alert Employees to Responsibilities

Most individuals engaged in research and development are justifiably proud of their accomplishments. Management should recognize this fact. At the same time, however, scientists and research engineers should be made aware of the fact that scientific information of great value is sometimes disclosed in meetings between scientists and engineering officials who are intent on letting others know of their achievements.

Management and security officials should inform employees that security depends on the continuing integrity of individuals who work with secret information. To emphasize this responsibility, some firms hold brief employee meetings in which a top member of management or the director of security answers questions and outlines the need for company security and points out that future job advancements and company benefits may be dependent on protection of company information.

Some firms ask employees to sign contracts to refrain from disclosing confidential information in the event the employee leaves the company. Basically, contracts of this kind fall into two classes. The first type is a signed agreement in which the employee promises to make no unauthorized disclosures of trade secrets in the future. In the second type of contract is a signed agreement in which the employee agrees to refrain from engaging in certain types of activities for another employer.

All courts are not in agreement as to the interpretations given to contracts signed by employees. Most courts, however, uphold a contract against a worker who has promised to refrain from disclosing trade secrets in the future. On the other hand, most courts look at contracts to ascertain whether the worker has agreed to bargain away his basic skills and right to make a living. When the court feels that this is the meaning of a contract, the courts will hold in favor of the worker. The basic legal doctrine here, then, is that the courts will protect trade secret information but they will not protect companies who attempt to prevent employees from making a living in the future.

Exit Interviews for Employees Working with Confidential Secrets

As pointed out in the chapter on personnel security, some businesses conduct termination interviews with all departing employees. If time limitations make such interviews generally impractical some firms still insist on conducting termination or exit interviews with all workers who have been engaged in trade-secret projects.

Usually, the company interviewer points out that the worker has been trusted and that this trust relationship is still valid. The interviewer sometimes makes a statement to the departing employee along the following lines: "We are satisfied that you have always been an honest employee and we wish you good fortune in your new endeavors. We are taking this opportunity to point out that you were trusted with confidential information and that it is our firm's right to retain this information. Since you have always been an honorable individual, we are satisfied that you will continue to be in the future."

The interviewer than may ask the employee whether all confidential work papers, notebooks, keys, blueprints or specifications, and other documents have been turned in. An appeal to the employee's integrity is almost always effective, and it places the employee on notice that the company is aware of the possible loss of trade-secret information and feels strongly about this possibility.

EMERGENCY READINESS AND DISASTER PROGRAMS

Many firms and institutions should include an emergency readiness or disaster program as a part of their basic security plan. This is to protect employees, facilities, and assets and to continue the ability to remain in operation under almost any circumstances. An effective preemergency program must incorporate strong, competent leadership from management, along with the organizational ability of the loss prevention or security department. Experience shows that emergency programs of this kind must be carefully planned and management must give strong support to the security official in charge of the program. Unless this planning and testing occurs prior to a disaster, employees may be thrown into panic. When panic occurs, employees and company property are almost always in serious danger.

In dealing with the unexpected, supervisors must be trained to first determine the nature of the emergency. Obviously, one set of plans cannot be used to deal with all unexpected situations. It is beyond the space limitations of this book to spell out complete details of an emergency readiness program for a business or institution, but some basic ideas can be set forth.

The supervising employee on the scene in any business or industry should be trained to go into action as soon as a disaster occurs, and to have employees ready to assist. A determination should be made quickly as to the nature, location, and seriousness of the program, as well as to whether it affects only the immediate area. An unexpected sewer stoppage is not likely to require the same attention as a fire or a bomb threat.

If the emergency appears to be serious, the supervising employee on the scene should immediately give an alarm to those around him and should make a call for help, both to management and to the indicated facilities such as the fire department or ambulance.

The supervising employee should then consider whether employees should be evacuated by predesignated safety routes and immediately direct efforts to minimize or combat the emergency.

For example, a leak in a chemical tank containing harmful ingredients could threaten the lives of company employees. The supervisor on the scene should do the following:

1. Evacuate threatened employees.
2. Obtain first-aid or assistance for anyone injured.
3. Alert employees in other departments who could be subsequently threatened.
4. Direct efforts to stop the leak.
5. Protect company assets or products that may be ruined or stolen if employees leave suddenly.

If there is no plan for protection of assets or products, for example, large amounts of cash could be exposed to theft in the event of a fire in a bank or in the cashier's office in a retail store. Preparation of a schematic drawing in advance of the problem will usually indicate specific areas where company assets could be exposed if employees leave on short notice.

In planning for emergencies, it is usually helpful to designate employees as alternates for the supervisor in charge at the scene. These trained alternates can fill in if the supervisor is on vacation, off shift, or on sick leave.

It is also vital for plans to include an individual who has been designated as the communicator. This employee has responsibility to immediately handle calls for the supervisor or leader. Without adequate communication facilities, an emergency plan can seldom be properly implemented, particularly where information must be transmitted to support facilities and members of management who can send assistance.

If the emergency is one in which a large number of people should be notified in a short time, communications should usually be transmitted via telephone using a pyramid system. In using this technique, one employee calls three or four others, who in turn call three or four others, on a prearranged schedule. In many situations, the communicator can utilize a walkie-talkie radio or other emergency transmitting systems. Radio broadcasts are not recommended, however, in situations where the emergency involves a bomb threat, since some types of radio transmissions can set off an explosive device.

There are some emergency situations that may render all facilities untenable, and it may be necessary to set up headquarters at an alternate site. Planning should consider this possibility, or if a disaster occurs in the computer room, for example, the remainder of the building may be usable but it may be necessary to have an alternate site for computer operations. This is discussed in some detail in the material on computer backup facilities.

Some disasters may, of course, occur at night or at another time when there are few employees in the building. If setting up an alternate site for the recovery program is necessary, it may be difficult to communicate with off-duty employees. It is therefore suggested that emergency plans include an up-to-date index of employee names, home addresses, and telephone numbers. If this index is available, employees can be quickly asked to report for emergency duty or to appear at an alternate site. A duplicate company payroll record may also be helpful. This is because some companies have discovered that employees "drift away" unless they can be notified about details of an emergency and can be instructed as to where to report for work in the future.

TELEPHONE CALLS, BOMB THREATS, AND BOMB PROBLEMS

In practically all instances, an anonymous individual who makes a call reporting that a bomb is set to go off in a specific installation does so for one of two reasons:

1. The individual making the call has definite information to the effect that an explosive or incendiary has been set and wants to reduce the possibility of personal injury or property damage. The caller may not necessarily be the individual who placed the bomb in the building, but may be someone working with the bomber.
2. The caller is a company employee or ex-employee who desires to cause problems or to get the day off if the building is evacuated.

Specific Instructions to the Telephone Switchboard

Valuable investigative leads can sometimes be obtained if the telephone switchboard operator is particularly alert and has had basic instruction in the handling of telephone bomb threat calls. Generally, more information can be obtained if the operator keeps the caller talking.

In San Francisco, a telephone operator recently asked the bomb threat caller, "Is the bomb in the main building or in the experimental laboratory across the street?" Since the firm had no building across the street and did not have an experimental laboratory in the state of California, management concluded that the call was a hoax when the caller said the bomb was in the lab.

A firm in the Chicago area allows many visitors to go through the company's facilities on tours. When a bomb threat call was received at this location, the telephone operator told the caller that a group of school children was scheduled to go on a tour through the facility in a few minutes. The telephone operator then pointed out that undoubtedly the bomber did not want to harm children and asked which area of the building should be made off limits for the children's tour.

It is a good practice to leave bomb threat telephone call report forms at the switchboard operator's station and to make certain that operators understand how the form is to be filled out. A typical form of this kind appears in Appendix 2.

Evacuation

It is frequently very costly to evacuate a building if the bomb threat is a hoax. This, of course, is a decision that must be made by top management. If the threat is real, employees may be killed or seriously injured unless the building is evacuated. Statistics show that by far the majority of bomb threats have turned out to be hoaxes, but this does not mean that calls of this kind can be ignored. If the building is evacuated, management has little choice but to search the entire building to ascertain whether a bomb has actually been placed on the property.

Keeping the Bomber Out of the Building

If the local police department or sheriff's office has a bomb disposal unit, the director of security should contact this unit to make certain the steps that should be taken in the event of a bomb threat.

Experience shows that good controls on means of access to the building may sometimes discourage individuals who have selected that particular business as a target for a bombing. This can sometimes be accomplished by strictly guarding entryways and by inspecting packages and materials entering critical areas.

Experience also shows that security and maintenance personnel should be trained to observe suspicious or unfamiliar persons carrying packages. When an individual of this kind is observed, the elevator operator or floor receptionist in a building may be able to keep the individual under control until a security representative arrives. If a security guard is available, the guard can keep the suspicious individual under observation until the suspect's intentions are clear.

For example, if the suspicious person loiters in the building without visiting anyone, the security guard should be trained to approach the individual and to offer assistance. If a suspicious individual declines this assistance, the security guard should ask the person to leave the premises.

If company receptionists are available in the building lobby or in the lobbies of specific floors, these receptionists can frequently assist in keeping out unwanted intruders. Usually these receptionists should have emergency call buttons in case they need assistance from a security guard or security representative.

Reporting the Location of a Bomb

If a potential bomb or explosive device is found in the search of the building, employees should be cautioned to take no action to remove or detonate this object under any circumstances. Handling should be left to explosives experts in the police department or bomb disposal unit.

If it is possibile to approach without moving or disturbing the suspected bomb, sand bags or mattresses can be placed around it. These should never come in contact with the suspected bomb, however. Steps should then be taken to block off the danger area with a clear zone of at least 300 feet, including floors above and below the suspected object. Windows and doors in the area should all be left open. This will usually minimize primary damage from an explosion and secondary damage from bomb fragmentation.

The building should then be evacuated and employees or other persons should not be allowed to enter the building until the device has been inspected and the bomb disposal unit gives permission.

The information here regarding discovery of a suspected bomb is merely an outline of steps that should be taken. Bomb detection, removal, and disposal comprise a complete subject that requires additional study by security professionals and is beyond the scope of an introductory look at security.

SUMMARY

Almost every business has information that would be of real value to competitors. Frequently, this information involves trade secrets: confidential processes or formulas, or methods for treating ingredients in the preparation of a saleable product. These trade secrets are not patentable and must therefore be protected.

American courts will issue injunctions to protect trade secrets or award damages to the plaintiff in a lawsuit against those who steal these materials. But before the courts will protect a trade secret, it must be proved that this information has been singled out and given special protection. Employees should be cautioned that such information must be kept secret and not revealed except in case of necessity. If a firm expects to obtain money damages or obtain an injunction from the courts, it must first show that reasonable precautions were taken to lock up written copies or otherwise safeguard the secret.

Almost any type of confidential information may be stolen by business spies: development projections, pricing schedules, new devices being developed, employee pay scales, client lists, research techniques, product specifications, and general know-how.

"Cloak and dagger" techniques are undoubtedly used by business spies from time to time, but relatively simple techniques are also used. Much can be learned by a competitor who pays a janitorial service for waste basket contents. Overly talkative salesmen and researchers often reveal too much. Purchasing agents and marketing employees who do not protect company records may also be instrumental in disclosing worthwhile information. Company scientists and research employees often reveal too much out of professional pride. In-house publications also may reveal too much.

Moonlighting employees and disloyal workers are also among the best sources of confidential information. An employee who leaves to join a competing firm can take with him the skills learned on the job, but legally cannot disclose confidential techniques or trade secrets.

If documents contain classified information, they should be locked up when not in use. Detailed classification, issuance, and lock up procedures may be justified in many instances. Control over document duplicating processes, over access to the building, and over destruction techniques are ways to safeguard confidential materials. Shredding machines are usually helpful.

Interviews of employees hired to work with confidential materials and exit interviews to brief departing workers are useful. An appeal to a worker's integrity is seldom misplaced.

Emergency readiness and disaster programs should be considered as part of the basic security program. Obviously, one set of plans cannot deal with all emergency situations. In many businesses or industrial establishments it is advisable to set up contingency plans for the types of emergencies that usually trouble that particular industry, as well a general plan for the unexpected.

Supervisors or managers on the scene should be taught to size up the emergency immediately, to call for help, to protect employees and property, and to take immediate action to control the emergency. This, of course, takes planning and teamwork.

Actual emergencies are almost always easier to cope with if employees have gone through simulated emergency situations in advance. Hard and fast rules cannot be applied to all problems.

Telephone operators should be trained to be alert for specific information in the event of a bomb threat call. They should note the caller's regional or foreign accent, any background noises, and any information that the caller will reveal. If the bomb threat appears to be real, management must decide whether to evacuate the building. Bomb squad officials should be notified of any bomb threat. If a bomb is located, evacuation should be immediate, with windows and doors left open. Handling of suspected bombs should be left to experts, and guards and security officials should confine their activities to evacuating the building and keeping out the curious until law enforcement officials or bomb squad members arrive.

REVIEW AND DISCUSSION

1. What is a trade secret?
2. Why do the courts usually accept the idea that trade secrets should be protected?
3. Describe the type of information that could be of interest to persons engaged in business espionage.
4. What are some of the less glamorous ways of obtaining confidential business materials or information as to a firm's plans and goals?
5. Outline some ideas to improve document protection in most companies or institutions.
6. Set out the basic steps that might be taken in setting up a general program against business espionage.
7. What are the company's rights with regard to trade secret information retained by a departing employee? Be as specific as possible.
8. What should be in a management program to alert employees to their responsibilities in protecting company information?
9. How may an exit interview be used in a program for the control of secrets?
10. Why does a company need an emergency readiness or disaster program?
11. Outline the most essential elements of an emergency plan for a typical firm or institution.
12. What options does management have when a bomb threat telephone call is received?
13. How can a telephone operator be trained to record clues that may lead to the caller's identity?
14. Should maintenance and janitorial personnel be trained to look out for intruders who could be planning to place an explosive charge or fire bomb in the building? What action should employees, elevator operators, receptionists, or guards take to keep intruders out?

15

Office Security Problems

The purpose of this chapter is to describe some of the security problems that may center around the office in a business or store. The acceptance of bad checks causes serious losses year after year in business. Some protective measures that can be taken are set out. Consideration is also given to protecting against loss of corporate valuables, office paintings, decorations and artwork, petty cash, business machines, and company equipment and supplies. Mailroom theft and security problems, payroll frauds, and other possible office losses are discussed.

BAD CHECKS

Business losses from bad checks were comparatively small prior to World War II. Since that time, fraudulent check-writing activity has increased at an alarming rate.

For a number of years, in excess of $2 million worth of bad checks has been reported annually by merchants in the city of Los Angeles. This figure is believed to represent only about one-half the actual amount of bad checks passed. Only a small percentage of the money lost through bad checks is ever recovered. From a security standpoint, then, it is far more effective to prevent the acceptance of a bad check than to attempt to recover on the loss.

Research shows that bad checks are passed most frequently in supermarkets, department stores, liquor stores, and gas stations. Banks usually account for only 5 to 7.5 percent of the fraudulent paper that is accepted.

There are seven types of bad checks:

1. Nonsufficient funds (NSF) checks, which are cashed by a customer with proper identification. Later the check is returned from a bank without payment because the customer did not have enough money in the account.
2. No-such-account (NSA) checks, drawn on accounts which either never existed or have been closed.
3. Incorrectly written checks, which have one or more serious mistakes that make the check unacceptable to the bank. Included in this category are postdated checks (dated after the time when the check was actually written), stale-date checks (more than 6 months old), or checks showing an amount in figures which does not correspond to the amount written out in longhand. Banks will regularly return checks of this type as unpaid, just as they will return a check that has been altered to cover the writer's mistake.
4. Stolen checks. These are checks that bear a forged endorsement on the back.
5. Deliberately altered third-party checks. These are usually U.S. Social Security or government checks, payroll checks, or cashier's checks which have been endorsed over to a retail merchant. The most common type of alteration is to raise the amount on the face of the check or altering the name of the person to whom the check is payable (the payee's name).
6. Fictitious checks bearing the name of a fictitious maker or payee.
7. Counterfeit checks which have been printed to resemble government or payroll checks but are on a paper that is slightly different from the proper check paper.

The most important security rule to follow in accepting checks is to require positive identification. Employees who accept checks in the business should be trained to establish positive identification. The basic principle involved is to accept checks only if the endorser or maker is known. Employees should be instructed not to accept any of the following for purposes of identification:

1. A temporary driver's license.
2. A Social Security card.
3. A selective service card, except to verify date of birth.

238 Office Security Problems

4. Receipts from companies such as the telephone company or the utility company for payment of bills.
5. A voter registration stub.

Some firms have found it helpful to have a rubber stamp made with which to stamp a form on the back of each check accepted. At the minimum, the form should require the following information: the name, address, and telephone number of the person cashing the check, as well as driver's license number, birthdate, physical description, and other identification offered. In addition, the employee accepting the check should make a notation "accepted by," followed by the employee's signature. Figure 30 is an example of the type of rubber stamp which may be used to record identification information.

Some firms have found photo-identity cameras effective in avoiding bad checks and other frauds. This type of security equipment records the face of the check passer and the document itself on a single frame of film. Although effective, this system is costly. A typical photograph of the check passer, the check, and the passer's identity cards is represented in Figure 31.

Some companies have experienced loss through failing to protect blank company checks and checkwriting machines properly. A business in Boston recently lost $18,000 because the company official signing checks did not carefully examine the company checks presented to him for signature. A bookkeeper at this firm regularly prepared checks for

FIGURE 30 Rubber stamp identification information.

```
NAME:_____
ADDRESS:_____
PHONE:_____
DR. LIC. NO._____
AGE:_____ BIRTH DATE_____
HT:_____ WT:_____
EYES:_____ HAIR:_____
EMPLOYEE:_____
OTHER I. D._____
1._____
2._____
3._____
```

Office Security Problems   239

FIGURE 31   A typical photograph of the check passer, the check, and the passer's identification.

Photo courtesy of Regiscope Corporation of America, New York, New York.

signature and placed them before the official. Although for many years the official had examined the supporting documents that accompanied each check to be signed, he eventually relaxed this policy and began to sign all checks presented without any verification. Taking advantage of this undue trust, the bookkeeper obtained the official's signature on a blank company check for $18,000. The bookkeeper disappeared shortly after cashing this check on a Friday afternoon.

PROTECTING CORPORATE VALUABLES

Sometimes business executives and security officers are so intent on protecting merchandise stocks and cash receipts that they overlook the security of corporate valuables. Blank (unissued) stock certificates, the corporate seal, corporate minutes, confidential board records, and company-owned stocks and bonds may all be deserving of protection. In a recent burglary of a New York City firm, the offender carried off the blank stock certificates, the corporate seal, some additional stock certificates that were being held in trust, and the stock issuance register. To date, the burglar has not issued fraudulent stock certificates from the stolen materials, but that possibility still exists.

In a situation of this kind, one of the problems may be that of reconstructing stock ownership records. From a security standpoint, the corporate secretary's records, seals, unissued stock and bond certificates, and other paraphernalia should be secured in a safe or vault. Large corporations generally use outside stock transfer agents, making these agents responsible for the protection and security of records. If stock ownership records are kept in the business office of a company, it may be advisable to maintain a duplicate record by copy machine or microfilm duplication in a safe outside location.

Recently, a Cincinnati corporation suffered a sizeable loss through embezzlement of company-owned negotiable common stocks. The securities in question were in the custody of the firm's corporate secretary, maintained in a bank safe deposit vault in the same building as the corporate offices. Six officials of this company had been given access to the safe deposit box, and it is apparent that one of these officials stole the stock certificates, since no one else had access. Eventually, the stock certificates were sold on the black market.

Many banks have agreements that allow access to a safe deposit box only if two company officials are physically present and sign for the box. A precaution of this kind fixes responsibility among company officials. The corporate security officer for the Cincinnati firm conducted all investigation possible. The president of the firm, however, refused to ask corporate officials to take a polygraph test to determine who could be responsible.

PROTECTING OFFICE ART AND DECORATIONS

In recent years, some firms have spent large sums to decorate executive suites and lobby areas to impress visitors and high-ranking callers. Sometimes it appears little thought has been given to the value represented by these furnishings.

The wife of a manufacturer in Los Angeles spent large sums of money assembling American Indian artifacts of unusual design. As the principal stockholder of this company, the manufacturer urged his wife to exhibit these art objects as wall decorations in the lobby, hallways, and executive offices of the firm. The pieces consisted of woven baskets of unusual design, rare Indian pottery, carved ceremonial masks, and Indian weapons. While no losses have been reported from this building to date, the company security man has pointed out that the building is invariably unlocked and unattended between 4 P.M. and 6 P.M. and that these artifacts could be stolen by anyone who wandered in.

The owner of a Manhattan company has become very wealthy by

constructing and leasing oil tankers and other merchant ships to private firms throughout the world. The owner of this company has a sentimental attachment to ship's models, displaying more than forty of these on the walls of the company offices. Practically all of these areas are accessible to the public during the day, and the entire office area is frequently left open by janitorial employees after 6 P.M. Collectors have pointed out that the value of these ship's models varies from approximately $2000 to $5000 each. Some of these items are irreplaceable. The owner of this company also displays a Rembrandt painting in one of the executive offices at this location.

In somewhat similar manner, a Maryland bank has displayed numerous antique items in the bank's lobby as well as other public areas. Some of these items are small and could easily be carried away. Because the antiques complete the colonial theme used in decorating the bank, however, they are evidently considered more as decorations than as valuables.

Security of art objects may be questionable unless they are placed in locations that are subject to control. Lobby receptionists, elevator operators, security guards, and other employees should be instructed as to the value of unusual objects and should be able to obtain assistance in protecting them if need be. If a firm wants to display a valuable item, it should consider displaying it in an inner office or in an area that can be locked.

Management and security representatives should also give consideration to whether art objects and high-value items are adequately covered by the blanket bond in the firm's insurance policies covering fire and theft.

The attorneys for a Dallas, Texas, firm found that many books in the company law library were missing. Investigation revealed that these attorneys were very friendly with other lawyers who maintained offices in the building and that these lawyers frequently carried off volumes from the company law library. A check with engineers in an engineering section of the same company revealed that many volumes from the firm's technical and engineering library were also missing. In situations of this kind, controls should be used to make certain that a book charge-out system is used if books are taken from the library.

OFFICE SUPPLIES

If office supplies are constantly pilfered, a company's operating costs may increase considerably.

There are a number of systems that may be helpful in preventing raids on company materials. Supplies can be maintained in a locked cabinet located in a central storeroom. Department and staff secretaries can then be permitted to requisition needed items once a week. Such a plan will enable a supervisor to review requests, employee by employee, to be sure that items requested are actually needed. Controls of this kind eliminate the problems that result when every employee in the office is given access to supplies. Unlimited opportunity is usually a principal factor in this kind of loss.

An alternate system that may be used is to permit each employee to requisition needed supplies on a mimeographed or printed requisition form at a specific hour each work day. This system eliminates the need for supervisory approval, as the quantities requested by each individual will be retained on file for a month or two. An occasional check of usage should be made by management to determine whether requests exceed needs.

PETTY CASH FUNDS

In one form or another, almost every business has one or more petty cash funds. These funds are used primarily to eliminate the need to write individual checks to pay for small purchases. Items paid for with petty cash generally include postage, railway express charges, small supply purchases, collect telegrams, and other business expenses below a set figure. The amount of money in a petty cash fund may range from $25 to $1000, according to the nature and the size of the business.

To set up a petty cash fund, management usually makes an estimate of those payments likely to be made over a relatively short period, usually not more than one month. An employee is then named as the petty cash cashier. Ideally, this employee should be bonded and should be an employee who is not assigned to the accounting department or one who has access to accounting records.

The fund is set up when the accounting department draws a check for an amount larger than estimated petty cash expenditures. This check is made payable to the cashier of the petty cash. Before expenditures are made by the cashier, another employee is designated as the petty cash officer, to approve and authorize expenditures from the petty cash fund.

When the need arises, the petty cash officer will accept documentation for expenses from the individual employee who has expended personal funds. A voucher is then prepared with one copy distributed to the employee receiving payment, the person designated to cash the petty cash check, and the petty cash officer.

After the cash disbursement has been made, the voucher is signed by the employee receiving the money. The cashier then records the transaction in the petty cash record and files the voucher with this documentation.

This process may continue until the petty cash fund is nearly exhausted. At that time, the cashier will summarize the expenditures for the period involved. Supported by the petty cash vouchers and receipts, the cashier will request replenishment of the petty cash fund from the auditing department. Prior to replenishing the fund, many companies will have the petty cash reconciled by a disinterested individual. This person will check all vouchers for the period involved. Working with the cashier of the petty cash, this person will inventory the money on hand and compare vouchers to documentation until the petty cash record is balanced. This balance will show the actual cash on hand and the amounts disbursed.

The type of fraud most frequently encountered in connection with petty cash funds usually involves alterations of vouchers. This can generally be avoided if vouchers are prepared in permanent ink or by typewriter, with the amounts of money written out in full. For example, a voucher representing an expenditure of $5 should be written "five dollars," not "$5."

It is usually regarded as a questionable security practice to allow company cash receipts or funds to be mingled with petty cash. When this happens, accounting may be difficult or impossible. Most firms utilize an unannounced audit of vouchers to make certain that embezzlements do not occur and that employees do not obtain cash advances from petty cash funds. Usually, the cashier of a petty cash fund is financially able to repay unauthorized advances. This is a form of embezzlement and should not be permitted. Experience shows that an employee who obtains unauthorized cash from a petty cash fund may also be involved in other schemes to steal from the company.

MAILROOM PROBLEMS

Using the Mailroom to Ship Stolen Items

A company that ships jewelry or any other merchandise of high value may expect losses because of thefts from U.S. mails if shipments are identified by contents. Accordingly, it may be wise to use a plain label and a return address that does not directly identify the shipper as a jeweler or other firm that would attract unusual attention. United States Postal inspectors do a good job in keeping mail thefts to a minimum.

However, a postal employee may place a preaddressed shipping label over the label that is already on a package. This preaddressed label will divert the shipment to the address of the thief or an accomplice working with the thief.

Theft problems may also be encountered in the company's mailroom. Unless there is adequate supervision, some company employees may send all their personal mail at company cost. A situation of this kind is usually revealed if management or security representatives make an unscheduled inspection of the outgoing mail to determine whether personal letters are being sent through company facilities. Some employees misuse company mail only at Christmas time, sending Christmas cards and packages to relatives through the company facilities. This type of misconduct should be prohibited by notifying employees in writing.

Most companies have encountered fewer theft problems by using a postage meter rather than retaining a supply of postage stamps. Experience in many companies shows that employees may regularly steal small quantities of stamps or pocket any change left in the mailroom to pay for postage stamps for personal letters. While these losses are usually small, the cumulative effect may be considerable. Use of a postage meter will eliminate this kind of theft to some extent.

Experience also shows that employees will embezzle by using the company postage meter, however, unless the meter is controlled and locked at all times when not in use. It is not uncommon for the key to the postage meter to be hidden in a location that is known to many employees, some of whom may misuse the meter.

Mailroom employees should be instructed that they are not to accept packages for mailing without management approval of each package. This is because employees in some firms will remove merchandise from the company warehouse and ship it to themselves or to a friend, using the company mail facilities to get stolen property out of the building.

Incoming Cash and Checks in the Mailroom

Many businesses receive a large number of checks or other payments through the U.S. mails. Usually a customer will not remit cash.

Whether the business is large or small, the process for receiving, opening, and distributing mail should not be delegated to one employee. Insurance companies report that embezzlement frequently results if only one employee is allowed to remove checks or cash from the incoming mail envelopes. When mail is opened by two employees, one should record the receipt of each remittance in a remittance log and the other should serve as witness.

Incoming checks should not be allowed to remain in the mailroom after receipt of the check has been recorded. Checks should be given to the employee responsible for posting and preparation of bank deposits. Insurance company experience shows that there should be a separation of responsibility in both opening the mail and posting.

In a recent case in Santa Monica, California, a hospital hired a mailroom clerk without background investigation. This clerk was assigned to open incoming mail, without another employee being present. After the clerk had been on the job for less than a week, a check was received for about $15,000. This represented a payment by an insurance company for a patient who had remained in the hospital for an extended period of time. The new clerical employee stole the check from the mail and destroyed the remittance letter. Opening a bank account in a nearby city, the clerk deposited the check and then drew out funds by writing small checks on the new account. This embezzlement was detected within a short time, when the insurance company did not receive a receipt for the incoming check. As soon as an inquiry was made by the insurance company, the new clerical employee at the hospital disappeared, taking with him the $15,000.

If a system is used in which two employees open all incoming mail, it is suggested that all checks be immediately stamped with a stamp bearing the notation "For deposit only to account of———." After this endorsement has been placed on a check, the loss in an embezzlement would fall on the bank cashing the check rather than on the company or institution that has placed the restrictive endorsement on it.

BUSINESS MACHINES

Many companies have found that pocket calculators, adding machines of all types, typewriters, and other business machines are frequently stolen and sold to pawn shops or other outlets.

If a file of serial numbers is maintained and a package pass system is used effectively, thieves will find it increasingly hazardous to remove business machines from office buildings. In addition, investigation of janitorial and maintenance employees, as well as supervisors, may be helpful in reducing losses of this kind.

Locking devices are available to secure typewriters and other pieces of equipment to desks and tables. Most of these devices are reasonably priced and effective in preventing theft. However, many of them destroy the mobility of equipment. Some stenographers dislike using a typewriter that cannot be moved about on a desk, or from desk to desk.

Some locking equipment uses a steel cord arrangement to provide a degree of mobility.

From time to time, businesses report the loss of expensive typewriters in "snatch and grab" thefts. In a case of this kind, the thief usually throws a rock or a brick through a glass door or window, grabs a typewriter, and runs away before anyone observes. Losses of this kind may be minimized if doors and windows are protected and if interior lighting is adequate.

Some firms have reported thefts of this kind at opening time, when only one employee is in the building and the employee has not relocked the front door. These incidents can usually be eliminated if employees and cleaning crews consistently lock office doors to prevent outside access.

PAYROLL FRAUDS

Employees in the office in many companies have responsibility for preparation of payroll records and for dissemination of paychecks or pay envelopes.

Some embezzlements are perpetrated by an employee who prints an extra paycheck and then cashes it. Office equipment is available that maintains an exact count of the number of checks that are prepared and signed on a check-writing machine. If audited regularly, the meter in this type of device will alert the auditing staff to the fact that an unauthorized check has been issued.

Recently, a Newark, New Jersey, firm found that a supervisor had placed such a "ghost" on the payroll. The supervisor was picking up this extra check every payday and had opened an account under another name in a nearby community for depositing the checks. Generally, this type of crime can be avoided if one employee in each section of the company has responsibility for preparation of the payroll while another verifies the hours worked. Another frequently used approach is to have a representative of the accounting department hand out individual paychecks only when an employee presents company identification.

Time card frauds are sometimes perpetrated unless the security representative or company supervisor observes employees to make certain that they punch their own time card rather than that of another employee.

TELEPHONE MISUSE

Unsupervised maintenance and janitorial employees frequently make long distance calls at company expense. This security problem may be eliminated at night if the switchboard can be locked off or supervised. If the switchboard cannot be controlled, small telephone locks are available to restrict the use of individual pieces of telephone equipment. These locks are usually available through telephone supply and equipment firms or through the local telephone company.

Security representatives in some firms recommend that the entire long distance telephone bill be audited, or at least that a reasonable percentage of the calls be spot-checked. If some of those listed on the telephone bill cannot be accounted for, a call to the listed number or a call to the telephone company business office will usually provide information as to who swindled the company.

SUMMARY

Losses resulting from bad checks increased greatly after World War II. In general, such losses can be controlled by requiring positive identification of the individual presenting the check. Business employees accepting checks should be taught that social security cards, selective service cards, and temporary driver's licenses may not be adequate for proper identification. Some businesses have found it helpful to use a rubber stamp form on the back of each check, with the person accepting making a notation on the form as to name, address, telephone number, driver's license, birthdate, physical description, and any reliable identification card. In addition, the employee accepting the check must be identified on the form.

Photo-identity cameras, incorporating a photograph of the passer, the face of the check, and the identification offered all on a single frame of film, have also been very helpful in reducing thefts through bad checks.

There may be times when the security of corporate valuables is overlooked. Blank (unissued) stock certificates, the corporate seal, company minutes, confidential board records, and company-owned stocks and bonds should all be protected.

Intent on a good public image, many firms spend considerable money for decorations and art work. Unless access controls are adequate, paintings and other art objects may be stolen. Technical libraries and law libraries may also suffer losses unless adequate charge-out and control systems are used.

Most firms do not maintain large petty cash funds. Amounts on receipts should be written out, so figures on the receipts cannot be easily "raised." While most thefts of petty cash by employees involve small amounts, activity of this kind could lead to larger thefts or embezzlements.

Unless the mailroom is properly controlled, employees may steal postage stamps or send personal letters and packages by unauthorized use of the postal meter. Confidential company documents may also be routed through the mailroom and may be subject to loss from this location. Incoming company checks may be pilfered in the mailroom. This can usually be prevented if two employees open incoming mail together.

Typewriters, adding machines, calculators, and other business machines are often stolen by thieves or outside burglars. Maintaining an inventory of machine serial numbers is usually helpful in proving theft, in establishing evidence in court, and in obtaining recovery of missing equipment. A number of security devices are available to lock typewriters and other machines to a desk, but most of these do not allow the kind of mobility that office workers frequently regard as essential.

REVIEW AND DISCUSSION

1. What is the most important security principle that should be followed to avoid losses from bad checks?
2. Why should protection be given to unissued stock certificates, corporate seals, corporate minutes and resolution books, as well as other corporate records? Explain.
3. Describe some precautions that can be taken to protect paintings that hang in an office lobby, art objects, or decorations.
4. Outline a workable system to control loss of office supplies.
5. What security measures would you suggest to protect a petty cash account from embezzlement?
6. What losses may be expected when there is little security in the mailroom? How do these losses occur?
7. What can be done to prevent the theft of business machines? To recover them after theft?
8. How can a "ghost" on a company payroll be detected? Amplify.

16

Special Security Problems of Institutions

Hospitals, schools, museums, and other types of institutions all may have unusual security problems. At the same time, institutional security programs will generally follow the same broad outlines that are used in business and other facilities, including security techniques described in other chapters. The purpose of this chapter is to outline some of these unusual security problems and how they may be handled.

BACKGROUND

Many institutions such as universities, schools, museums, and hospitals are complex business organizations. Facilities of this kind are often small communities in themselves, with the varied responsibilities and demands that are found in any city.

There is almost always an air of openness connected with these institutions, since almost all of them were founded to assist the general public. In the minds of some, these institutions' worthy objectives seem to be opposed to any restrictive rules and regulations, especially those that are a necessary part of a working security program. In actual practice, however, there must usually be a blending of the two extremes for the good of the institution.

To outline security objectives, it may be well to examine some of the problems faced by these organizations. A large hospital, for example, may be located in five or six buildings with a floor area of 7 or 8 acres. The payroll of an institution of this kind will probably exceed 1000 employees per year. The institution's laundry facilities may handle approximately 24 or 25 tons of patient gowns, towels, bedding, table linens, and similar items in a week. The cafeteria or kitchen in an organization of this kind may produce approximately 2500 meals per day for patients, employees, and visitors. A public university may have an even greater number of buildings and employees than a hospital.

HOSPITALS

Many businesses are able to close their doors to the public for a good part of each day. A hospital, however, is a public building that usually remains open and in operation seven days a week, with facilities available to visitors and with many activities taking place 24 hours per day. A hospital cannot be simply locked up and ignored for any length of time. In addition, patients are often comatose or unable to take care of themselves and the hospital must provide protection to the young, the old, the sick, and the helpless.

To the majority of the public, hospitals are still a place where injured and sick individuals are cared for—institutions outside the climate of distress and crime. But in recent years, criminals have attempted to take advantage of the availablility of dangerous drugs and narcotics and have preyed on helpless patients unable to protect their personal belongings.

Then, too, in a hospital just as in any large institution, stockrooms, supply areas, and kitchen pantries are susceptible to wide-scale theft unless controls are utilized.

Parking and Escort Problems

Parking is frequently a serious problem for a hospital. There is sometimes a shortage of space for both visitors and employees who must be accommodated. Orderly control of traffic may be of help in permitting a maximum number of visitors to spend time with patients.

Obtaining help by way of a parking lot guard's walkie-talkie radio may mean the difference between life and death for the arriving victim of a heart attack. Guards can also be of considerable assistance to the emergency room nursing staff in helping an injured or critically ill patient into the emergency room.

Security coverage of the hospital's parking lot may be critical because thieves can be counted on to break into a doctor's car if they can observe the doctor's bag sitting on the seat, even though the car is locked. The director of security can frequently prevent such thefts by teaching doctors to lock medical bags in car trunks or to carry them into the hospital. In thefts of this kind, the object is usually to obtain narcotics from the doctor's bag. Since many hospital employees work throughout the night, batteries, tires, and other accessories may be stolen from employees' cars unless security patrols are effective.

Then, too, there are usually large numbers of women on a hospital staff. If there is a nurses' residence located near the hospital, there is always the possibility of a purse theft, an assault, or a rape. Historically, hospitals have always attracted peeping toms and sex criminals.

The hospital security staff is called upon regularly to conduct investigations of crimes on the premises. These are usually situations involving thefts, but a wide range of other crimes occur too. The guard force is frequently called upon to assist floor nurses, especially in situations where a muscular, strong patient is experiencing a medical reaction and is threatening visitors or hospital staff, or destroying property. Armed robbery is always a possibility in a hospital neighborhood, especially since many hospitals accumulate large sums of cash in the check-out process. The security staff is frequently called upon to escort the hospital cashier, the cafeteria or coffee shop cashier, or the individuals in charge of the gift shop who need protection against robbery.

In a number of hospitals, the security department has responsibility for supervising lost and found property and for returning items that are turned in.

Access Controls

There is an almost continual flow of traffic into and out of a hospital. Few patients remain for any appreciable amount of time, and visitors also change constantly. Some hospitals utilize passes to control visitors, but others feel that there is no need for such passes during scheduled visiting hours. Control systems are usually needed to keep out individuals who are not working at the hospital and who are not legitimate visitors. Badges to identify hospital personnel are helpful in this regard. If visitor passes are issued, they are usually more effective if color-coded by floor or department to be visited.

Theft Problems

Employee theft is one of the basic problems in present-day hospital security. Narcotics, drugs, and vitamins, for example, are always subject to theft unless adequate procedures are set up and followed for issuance of these substances.

In a hospital in Dallas, Texas, it was recently found that the head pharmacist had embezzled large amounts of narcotics by forging narcotics charge-out slips for patients, some of whom had died prior to the time that the fictitious charge-out slip was prepared. Foodstuffs from a hospital cafeteria or coffee shop, institutional silverware, and supplies of all kinds are frequently stolen.

In hospital laundry facilities the replacement costs are frequently staggering, owing to the fact that employees consistently steal sheets, pillow cases, and towels. Patients also are responsible for linen losses, wrapping shoes or clothing in hospital towels or using pillowcases as laundry bags.

Receiving procedures in institutions are frequently lax, with the receiving clerk signing for deliveries that are never counted. When drivers observe a situation of this kind, it can be expected that they may consistently short the count of merchandise left on the receiving dock. The security department can frequently eliminate problems of this kind by making spot-checks of deliveries at the time they are left on the receiving dock.

Package Passes

Employees at many hospitals are allowed to take out packages only when given a package pass; the pass is collected by a guard as the worker exits. To be an effective deterrent to theft, package passes should be verified according to the principles set out in the material on package passes.

Experience also indicates that incidence of theft can be reduced if employees entering and leaving a hopsital are required to use an employees' entrance and to show identification cards.

In the hospital food service area, adequate locker room facilities should be available and frequent spot-checks should be made of lockers both for health and for security reasons. Unauthorized individuals should not be permitted to loiter in these areas and guards should give special attention to locks on refrigerator doors where foodstuffs are stored, to make certain that hinges are not unscrewed so that doors can be easily removed.

Thefts from Patients

Thefts from patients are frequently a troublesome problem in hospitals. The admitting form used by the hospital should require a patient to read a statement cautioning the patient to deposit all valuables in the hospital housekeeping envelope and explaining why. One hospital in the Los Angeles area that has been successful in minimizing thefts from patients requires the director of security to monitor admitting procedures from time to time and to report to management whether admitting clerks are spending sufficient time in cautioning patients to protect their valuables.

Payment for Hospital Medication and Services

Few hospitals could function for long without the revenues obtained from regular charges for services and medications. Whether bills are paid is usually regarded as an administrative problem in the hospital. However, a hospital's director of security may be asked to make spot-checks of patient charges to insure that medications and services are actually billed to patients. A periodic check of departmental records against the records maintained for individual patients will usually disclose whether billing is being done properly.

Disaster and Evacuation Plans

Hospitals are particularly vulnerable to all types of disaster because many of the patients are in such physical condition that they cannot assist themselves in the event of an unexpected emergency. If the building is very large even mobile patients will be unlikely to know the physical layout of the building.

In addition, most hospitals stock a number of flammable chemicals, and hospitals almost invariably accumulate large amounts of flammable trash. Large bottles of oxygen may also add to the hazard. Most authorities on fire protection are in agreement that every hospital should have a detailed fire prevention and evacuation plan. Usually, the plan should be drawn up and handled by a fire marshal who is an employee normally charged with other job responsibilities. This fire marshal should report to the director of security and safety.

Every employee should be trained to perform specific duties in case of fire and to carry these duties out under the direction of floor captains. Fire training must be repeated over and over even though such repetitiveness may induce employees to become lax through boredom. It is the responsibility of the fire marshal to have regular drills to make certain that employees perform as expected. He must make every effort to avoid letting the program for fire training become too routine; if employees regard their duties as a mere nuisance, the safety of patients may be in jeopardy.

Emergency plans should also cover evacuation in the event of other disasters. Medical advice must be sought for planning how to handle emergency support systems and intensive care patients. Planners should also consider possible alternative power sources if the electricity that operates critical equipment and machinery in intensive care units fails.

SCHOOL SECURITY

In recent years, there has been a serious increase in vandalism of schools and school property. This may be the most serious problem facing security directors in public school systems.

The causes of vandalism are various. The following conditions are usually involved in this type of senseless destruction:

1. The breakdown of the family group and family control.
2. Drug abuse.
3. Street crime by students and individuals hanging around a school.
4. The glamorization of violence in movies and on television.
5. Anonymity in the neighborhood and in the schools and the prevalent attitude that the actions of individuals are of little concern to others.
6. Diminished public concern for the value of property, both personal and public.
7. Permissive attitudes in the schools toward student behavior.
8. Permissiveness in the courts with regard to juvenile offenders.

When apprehended following acts of vandalism, the culprits are seldom able to give any real reason for their antisocial activities. Some become involved merely to have something to do.

The cost of protection against vandalism and violence is certainly very high, but no exact figures are available. Part of the reason for this is that the resources and experience of the security industry as a whole have only recently been brought to bear on the problem.

If a school is used after school hours by outside organizations and groups, this seems to encourage the community to take an interest in the school system and to protect the buildings against vandalism. Some security officials who have had experience in this field recommend that architects begin to use vandal-proof materials both to build new structures and to repair old ones. Security glass can be installed in some windows, and windows and light bulbs can be placed as far out of reach as possible.

General apathy on the part of the public appears to be one of the factors that contributes to vandalism. If adults can be educated to keep an eye on the schools and immediately report the presence of loiterers or suspicious individuals, vandalism will decrease. Some schools and institutions in England have had success by maintaining so-called live-in personnel, teachers or administrative officials who maintain a residence on the school property.

In the United States, some authorities report that student patrols have been helpful in augmenting the efforts of faculty, security, and police personnel. Alarm systems, installed in selected buildings, have proved effective in helping to apprehend vandals and in reducing nighttime and weekend vandalism.

Success has also been achieved where communities have been able to appropriate sufficient money for the proper training of school security officers. These officers are most effective if they are trained especially to deal with adolescents. In Detroit, Michigan, for example, the regular city police academy has a program for training school security officers.

School officials also have found it helpful to keep the courts informed of the background and activities of students arrested as vandals. Whereas some students can be helped by counseling alone, other students and loiterers around the schools may continue to be involved in vandalism until actually sentenced to jail.

Educational programs run by school staff officers have also proved effective. Usually these are directed toward convincing individual students that they have a vested interest in curbing vandalism for the good of their school and the taxpayers in the community. Some school boards pay a reward to students who turn in an individual guilty of vandalism.

Some school districts have found alternative educational programs helpful in turning problem students from vandalism to more profitable activities. Programs of this kind have been successful in Ann Arbor, Michigan, Philadelphia, Pennsylvania, as well as other locations.

Student identification cards can be valuable aids in controlling vandalism and other crimes on the campus. If students are required to carry valid identification, they are less likely to engage in misconduct. A numbered identification card with the student's photo on it can also be used in the school library to check out a book. A library scanner reads the number of the identification card and relays this information into the library computer. This enables the library to keep track of charge-outs and delinquent books.

In recent years, crimes of all kinds have increased around the schools. Public school administrators are responsible for protecting students, teachers, and property as well as for educating students. It is important for school authorities and security officers to be aware of the legal limitations of the school security system. To some extent the authority of security officers is determined by the status of these officers as sworn law enforcement personnel, special police officers, or civilian guards. Often, however, local court decisions or statutes put specific limitations on the scope of actions these officers may take. School security officers thus must be clearly informed as to the policies that should be followed and the laws that protect both students and school officials. In recent years, the National Association of School Security Directors (NASSD) and the Law Enforcement Assistance Administration (LEAA) have both been helpful in providing training courses for school security officers and administrators.

Additional Problems on College Campuses

In addition to the security problems encountered in the public schools, major universities and colleges must meet other challenges. The size of such campuses alone may seriously hamper communications and policing functions. The main campus of the University of Houston, for example, is located on a tract of 390 acres.

Thirteen strategically located emergency call boxes and a powerful radio station and repeater-booster located on the roof of the administration building at the University of Houston helped considerably in solving the communications problem. Call boxes were placed in unlighted, heavily wooded sections of the campus. Boxes were positioned so that any student on campus would never be more than 4 minutes away from radio contact with the campus police station.

A college campus is very difficult to patrol on foot; accordingly, automobiles or Cushman vehicles are almost always necessary. Experience shows that a Cushman vehicle offers approximately four times as much patrol coverage as can be done on foot. It should be pointed out, however, that the Cushman patrol car does not have the speed for a fast response to an emergency, and automobiles should be kept available.

Equipping patrolmen with walkie-talkie radios is a distinct advantage on a university campus. When the patrolman needs help, needs orders, or observes any situation worth reporting, information can be passed on immediately by radio.

The policing capability of a university or college security organization may be pushed to the utmost when there are special events on campus, during football games or other sporting events, or at registration. Special planning and training is necessary for these periods.

Theft of institutional property and equipment has become increasingly troublesome in recent years. In a business, management usually does not hesitate to require employees to lock offices and protect machines and equipment. On a university campus, the security department frequently encounters difficulty in obtaining compliance with requirements to lock offices and protect property. In many instances, security officers simply are not given the tools to control theft. Backing by administration is essential for campus police effectiveness.

Other crimes on campus may also be a serious problem. Some areas have been plagued by rapes, sexual assaults, and student robberies on campus. Incidents of this kind require close coordination with regular police agencies, long hours on patrol assignments, and preventive educational programs for students. To handle such crimes, the campus security force must be trained to make arrests properly, to recognize and protect evidence, and to supervise and handle search procedures. In spite of outbreaks of campus crime, there seem always to be some elements on the campus who resent even essential policing activities.

Muggings and armed robberies are ever-present possibilities on large, wooded campuses, especially where students may be on the move at all hours of the night. Robbery of campus funds at registration time is also a serious threat, since enormous amounts of cash may be on hand in the registrar's office.

Very controversial speakers may appear to lecture at any college or university, sometimes in an atmosphere of near-riot, and this adds another dimension to campus security. At times, a speaker may need to be protected by an escort of guards, and at other times, the speaker may incite students to riot, in which case campus security must maintain order.

Some of the facilities of the university or college may warrant special protective programs. The campus computer facility is a sensitive area of this type, having been the target of activists and antiwar groups in the past. It is hoped that this type of campus activity has come to a halt, but the security department at any college or university is usually in need of updated contingency plans for emergencies.

In addition, a campus security department usually bears some responsibility for fire safety and OSHA[1] safety programs.

---

[1] Occupational Safety and Health Act. See Chapter 17.

MUSEUM, LIBRARY, AND PARK SECURITY

Museums, libraries, and public parks all have security problems that are common to other institutions as well as their own individual problems. Museums all over the world have been seriously troubled by the theft of priceless art treasures. Alarm systems are frequently necessary to protect each individual piece of art. Communication between security officers also poses a serious problem in many museums, as it is essential they remain in constant communication with each other and with museum headquarters.

The National Endowment for the Arts, an independent governmental agency, was created in 1965 to encourage and assist cultural endeavors throughout the United States. Under certain conditions, this endowment provides museums with funds for the renovation and improvement of museum security, climate control, and storage. Funds from this source have been helpful in upgrading security in some institutions.

Theft and fire problems and how to protect rare manuscripts are among the troublesome security situations that face libraries.

Private parks may handle as many as 30,000 visitors in a day. As one park security director has put it, "A few of these people forget to leave their troubles at home." As a consequence, security staff members at some parks are being trained through courses at the local police academy. Injuries and illnesses are frequent in parks, and it is usually helpful for security officers to be trained in first-aid and cardiopulmonary resuscitation and to qualify as paramedics.

Security protection is frequently needed at sports events, such as the U.S. Open Golf Tournament. Some major security companies can provide complete security and control programs for conventions, for exhibits, for major sports shows, for traveling exhibitions, or for a circus.

SUMMARY

Hospitals, schools, and all types of public and private institutions have both common and specialized security problems. While some institutional problems are unique, most of the security techniques described in this book can be brought to bear on them.

Security problems in these organizations often arise from the fact that institutions are frequently complex businesses—they are often cities in miniature. But unlike most conventional businesses, institutions are usually devoted to the public good and are unable to exclude the public from many of their activities. A hospital, for example, usually admits visitors 7 days a week. Then, too, many of the patients are comatose or unable to take care of themselves. Protection must also be extended to nursing and medical staff members and patients who are very young or very old.

Uniformed guards are vital in assisting hospital visitors, administering to heart attack victims, and getting critically ill patients into the building. They keep narcotic addicts from breaking into parked cars owned by doctors and help prevent attacks or robberies, especially of employees who work the late shift.

Access control systems are often vital to hospital security, as visitors must be accommodated but intruders kept out. Thefts may be very frequent in some hospitals unless secure controls are kept over foodstuffs, supplies, and equipment. Narcotic supplies and patient funds are also targets for thieves. Guard coverage and package pass systems may be very effective in eliminating this kind of theft.

Because of the helpless state of many patients, it is imperative that hospitals have workable disaster and emergency evacuation plans. Fire training is a continuing responsibility in almost all institutions.

Schools frequently experience loss through arson, burglary, and student vandalism. This needless destruction of property has reached great proportions in recent years. Generally, vandalism has been attributed to a disintegration of family life and community values.

Alarm systems and uniformed guards have helped when available, but only at great cost. Use of vandal-proof materials in construction or repairs has been helpful, including masonry construction, break-resistant windows, and durable lighting fixtures.

Success has been achieved in some areas by appropriating sufficient money to train school security officers. In many communities it has been found that vandals will repeat their activities until school and police authorities persist in complaints to judges and a jail term results.

Education of school staff officers has also proved helpful, with programs directed toward convincing individual students that they have a vested interest in stopping vandalism, for the school, for the community, and for their parents as taxpayers. Some school districts have found it effective to pay rewards for information about vandals. Good student identification programs, to pinpoint nonstudents on the campus, have also paid dividends.

Museums and other institutions often must rely heavily on an effective guard force, on good perimeter controls, and on sophisticated alarm systems.

REVIEW AND DISCUSSION

1. What are the factors that make it difficult to keep potential security violators out of hospitals, schools, or institutions?
2. Granted these institutions have unusual security problems, may they still make use of many protective techniques that have been discussed in this book?
3. Why is the parking lot a particularly troublesome area at a hospital?
4. Explain in detail why it is usually desirable to escort nurses and employees to their private vehicles or to nurses' quarters?
5. Why is it helpful to check receiving of supplies and goods delivered to the hospital dock or stockroom?
6. Explain why emergency evacuation of a hospital may be a very critical operation.
7. List some of the major factors or conditions that lead to vandalism in the public schools.
8. What are some of the techniques and devices that have proved helpful in reducing vandalism and arson at school facilities?
9. Explain how communications problems have been solved on some university campuses and in museums.
10. Do private security companies contract to provide security coverage to sporting events and shows?

17

The Occupational Safety and Health Act

In recent years, farsighted security and management officials have come to the realization that a preventive approach eventually reduces security losses and saves money. Almost invariably more loss can be prevented than can be corrected by investigation conducted after the fact. This is the so-called loss prevention approach to security. Most businesses are also developing a preventive approach to safety. Until recent years, however, security and safety were often viewed as unrelated areas. Because of the investigative aspect of security, some companies have regarded security as a punitive or investigative function. On the other hand, safety has usually been associated with workmen's compensation problems, injury losses, and claims by personnel.

Many of the objectives of both loss prevention programs and safety programs are identical. Both programs affect employee production, morale, and integrity and are heavily committed to the protection of company assets. As the loss prevention philosophy has developed in security, more and more companies have combined the responsibilities for security and safety under a company director of loss prevention.

Among the basic problems of loss prevention is compliance with the Federal Occupational Safety and Health Act of 1970 (commonly known as OSHA). At the time of the industrial revolution, around 1800, machinery came into common usage in England and the United States.

Hours were long and working conditions were often dangerous and unsatisfactory.

Gradually, the social conscience brought public opinion to bear on problems in sweat shops and factories. But changes in public thinking came slowly. Eventually, the courts began to hold managers and property owners responsible for dangerous conditions and for negligence in failing to provide decent places to work. A workmen's compensation law was passed in Germany in 1883. Similar laws followed in most of the European countries by 1900.

There was long opposition to workmen's compensation laws in the United States. Objections were usually based on the fear of increased production costs and the possibility that workers would fake injuries and malinger on the job. In 1902, Maryland became the first state to pass a law of this kind, but between 1910 and 1920, 42 states and three U.S. territories passed workmen's compensation laws. By 1940, statutes of this kind were in effect in 47 of the existing 48 states.

As a result of the adoption of workmen's compensation laws, accidental deaths and injuries decreased considerably. But by 1958, this trend had come to a stop. By 1968, injuries and deaths from industrial accidents had begun to rise. As a result of these increases, the Williams-Steiger Federal Occupational Safety and Health Act of 1970 (Public Law 91-596) went into effect on April 28, 1971.

PURPOSE AND BROAD COVERAGE OF THE LAW

The Occupational Safety and Health Act of 1970 (OSHA) is very broad in scope, applying to almost every factory, warehouse, or industry, as well as many retail establishments that are engaged in business affecting interstate commerce. An estimated 61 million U.S. workers in more than 5 million industries and businesses are covered by the law. Applied in all 50 states, the law also pertains to the District of Columbia, Puerto Rico, the Virgin Islands, American Samoa, Guam, Wake Island, and other federal territories.

The stated purpose of the law is "to assure so far as possible every working man and woman in the Nation safe and healthful working conditions and to preserve our human resources."

The terms of OSHA do not apply to situations covered by other federal safety or health laws, such as those under the Federal Coal Mine Health and Safety Act or the Atomic Energy Act of 1954. Federal, state, and local governmental employees are specifically excluded from coverage.

RESPONSIBILITY OF EMPLOYERS UNDER OSHA

Every employer in a business or industry that affects interstate commerce has the responsibility to furnish each employee a place of work that is free from recognized hazards or conditions likely to cause serious physical harm or death, and the employer must maintain the premises to comply with these conditions. In addition, individual employees have the duty to comply with those safety and health standards, rules, regulations, and orders issued pursuant to the act that are applicable to the employee's individual actions and conduct.

ADMINISTRATION OF OSHA

The enforcement and administration of OSHA are handled primarily by the secretary of labor and a new agency, the Occupational Safety and Health Review Commission, set up with three members appointed by the president.

Research about safety and health conditions is handled by the secretary of health, education and welfare (HEW) whose functions are usually handled by the newly created National Institute for Occupational Safety and Health within HEW.

The U.S. secretary of labor is responsible for inspections and enforcing job safety and health standards, and inspections are made from offices located throughout the country.

In general, the rules under OSHA govern conditions which have been proven by experience and research to be likely to cause harm or injury. Several thousand standards have been set up to cover employers and employees. While many of these standards pertain to specific industries or types of workers, most of the standards are those generally met by companies that practice good safety and provide good employee protection.

For example, one rule requires that passageways and aisles in industrial establishments and warehouses be kept in good repair and remain unobstructed. Another example is the rule requiring employees to use suitable face shields or goggles if engaged in drilling, riveting, grinding, pouring metal, or any activity in which there is a hazard or potential hazard to the eyes.

It is the responsibility of all employers and workers to familiarize themselves with the requirements and standards that apply to their specific work area and to observe these requirements at all times. In addition, the secretary of labor has the power to add rules or to modify, revise, or revoke existing rules, on the basis of information submitted to him by the secretary of health, education and welfare.

If an employer or employee feels that one of the OSHA rules is an improper restriction, the employer or employee may challenge the validity of the rule by petitioning the U.S. Court of Appeals within 60 days after the new rule is put into effect. Emergency temporary rules can be set up and become effective if published in the Federal Register.

To prevent injustices to businesses, the act allows the secretary of labor to set up a hearing on an employer's application requesting temporary variances from the OSHA standards. A variance is sometimes necessary to give the employer sufficient time to comply and is often granted if the employer can show need for a time extension and presents an interim plan for dealing with the problem. Affected employees can also request a hearing in the event they feel that a new rule on working conditions has been improperly imposed.

An OSHA reference guide specifies the following as typical standards required under the act:

1. No employee dealing with toxic materials or harmful physical agents will suffer material impairment of health or functional capacity, even if such employee has regular exposure to the hazard dealt with by such standard for the period of his working life.
2. Development and prescription of labels or other appropriate forms of warning so that employees are made aware of all hazards to which they are exposed.
3. Prescription of suitable protective equipment.
4. Monitoring or measuring employee exposure to hazards at such locations and intervals and in such manner as may be necessary for the protection of employees.
5. Prescription of the type and frequency of medical examinations or other tests for employees exposed to health hazards. At the request of an employee, the examination or test results shall be furnished to his physician.[1]

OSHA VIOLATIONS

If a worker or an authorized representative of a worker believes that a safety or health violation exists, or that imminent danger threatens, the employee may request an inspection by OSHA. This is done by sending

[1] U. S. Department of Labor, Occupational Safety and Health Administration, *A Handy Reference Guide—The Williams-Steiger Occupational Safety and Health Act of 1970* (Washington: Government Printing Office, 1971) pp. 4–5.

a written notice to a representative of the Department of Labor or to an OSHA Office.

In enforcing the act, OSHA safety inspectors may enter without delay and at any reasonable time. The act allows the employer and a representative authorized by company employees to go with the inspector during the physical inspection of any work place to aid and inform in the inspection. In making inspections and investigations, the OSHA inspector (under authority of the secretary of labor) also has power to require the attendance and testimony of witnesses who have firsthand knowledge and who can produce evidence bearing on a possibly dangerous condition. The OSHA inspector is authorized to question both employers and employees to get at the facts.

If no reasonable grounds for a complaint are found, the OSHA inspector will notify the complainants in writing.

If the inspection reveals a violation of standards, the employer is given a written citation, describing the specific nature of the violation. All citations are set up to provide a reasonable time to correct the violation. If the violation is minimal or merely technical—that is, where the situation has no direct or immediate relationship to safety or health—the inspector may issue a notice rather than a citation. Citations are not issued for violations that occurred more than 6 months previously. The real intent of the law is to obtain good compliance from employers. However, the penalties for failure to respond can be very effective. Within a reasonable time after issuance of a citation for a violation, the employer is notified by certified mail of the penalty OSHA intends to assess. The employer then has 15 working days within which to notify the Department of Labor that the company desires to contest the citation or proposed penalty. If the employer fails to notify the department within the 15-day period, the citation and assessment are final, under the terms of the act.

The employer is given a hearing if the citation is contested. After hearing both sides, the commission will issue an order either affirming, modifying, or doing away with the citation or proposed penalty. If the employer is not satisfied with the ruling, the employer may appeal to the U.S. Court of Appeals.

TIME FOR CORRECTION OF HAZARDS

Obviously, some existing hazards cannot be corrected overnight without creating serious problems for the employer. When OSHA issues a citation, the citation states a reasonable time for elimination of or protection against the hazard. An employer may contest this time limit if he notifies the Department of Labor within 15 days. The time limit is then reviewed by a Department of Labor review commission. Additional time will be granted to the employer if the request appears to be reasonable and made in good faith.

If the violation is not corrected within the time approved by the review commission, the employer is at fault and is subject to monetary penalties of up to $10,000 for each violation.

Serious violations are defined as those creating conditions under which there is a substantial probability that death or serious harm may occur. Citations issued for this kind of violation carry mandatory penalties of up to $1000 for each violation. An employer who continues in failing to correct a violation for which a citation has been issued may be penalized up to $1000 each day that the violation continues to exist.

A willful violation by a company or employer that results in the death of an employee is punishable by a fine up to $10,000 or imprisonment of up to 6 months. A second conviction carries a penalty of double the monetary fine and jail term. In addition, it is a criminal violation to make false statements or to give unauthorized advance notice of any OSHA Inspection.

RECORD-KEEPING REQUIREMENTS

One of the basic purposes of OSHA is to find out how and why serious accidents occur. When this kind of information is available, OSHA regulations can be modified or reviewed to make certain that serious accidents are reduced. Accordingly, there are some record-keeping requirements under the act.

Employers who are involved are required to keep and make available records on work-related deaths, injuries, and illnesses. Minor injuries requiring only a minimum of first-aid and treatment need not be recorded. But a record must be maintained of each incident involving medical treatment, loss of consciousness, restriction of work or employee motion, or transfer to another job.

Under the act, businesses can also be required to maintain accurate records of employee exposure to potentially harmful chemicals or toxic materials and to advise employees promptly of excessive exposure and of any corrective action being taken. Specific OSHA forms are required for the maintenance of records. OSHA Form 100 is a log that must be filled out at each individual work establishment concerning any serious injury, illness, or death.

Another form, OSHA Form 101, is a supplemental record of details concerning individual illnesses, injuries, or incidents. Both OSHA Form 100 or Form 101 must be filled out in detail by the employer within 6 working days after notification of a specific incident.

OSHA Form 102 is a summary prepared on an annual basis by the employer, setting out information in Form 100 and Form 101. This form must be maintained in a place where employee notices are normally posted, must be in place by February 1 of each year, and must be kept on display for at least 30 consecutive calendar days. The posting of OSHA Form 102 is in effect a statement as to the effectiveness of and problems in the safety program at the local level. This information is usually helpful to both employees and the employer in correcting problem areas.

An additional form, Form 103, is required of employers of more than 100 persons and must be completed and returned to the Department of Labor.

THE EXTENSIVE NUMBER OF OSHA STANDARDS

There are too many specific standards required by OSHA to attempt to list them all here. The loss prevention director of a company can obtain copies of OSHA requirements from the nearest office of that federal agency.

Many employers have opposed OSHA requirements, protesting that compliance in all areas would require the expenditure of large amounts of money to modernize manufacturing or industrial plants. On the whole, however, the requirements are consistent with basic standards of health and safety and there is little question about the long-term benefits that will be obtained from the act. Some employers have withheld compliance on the grounds that constant regulation of private business should not be allowed if individual freedoms are to prevail in this country.

Although many OSHA requirements apply only to specific industries or types of business, others are generally applicable. For example, OSHA standards require buildings designed for employee occupancy to be furnished with sufficient exits to permit prompt escape in the event of fire or other disaster. An adequate supply of drinking water must be provided and individual drinking facilities furnished. Specific sanitation standards must be met in every working establishment. The act further specifies that sanitation requirements be provided according to the number of employees working for the company, both with regard to washing and toilet facilities. If employees are permitted to eat on the premises, the areas used must be maintained in a sanitary condition, away from toxic chemicals or dangerous work areas.

Fire extinguishers and protection facilities must meet minimum standards and scaffolds or dangerous work areas must be properly maintained and supported.

Aisles in warehouses, stockrooms, and industrial areas must be maintained in a safe condition, and open vats, holes, tanks, and pits must be protected by guard rails or covers that meet OSHA specifications.

These are only a few general requirements. The director or manager of loss prevention should study the requirements for each specific facility in detail. Other security employees, supervisors, and guards should be educated to check and inspect areas where OSHA violations are likely to occur. Company officials should be notified immediately by written report if violations or potential violations are observed. Top management in the company should retain the right to make decisions regarding OSHA regulations, but decisions cannot be made properly unless management is advised of all violations or potential trouble areas.

SUMMARY

The loss prevention approach is effective in most companies in reducing both security and safety losses. The Federal Occupational Safety and Health Act (OSHA), effective April 28, 1971, was passed to provide safe and healthful working conditions in every business or industry that affects interstate commerce.

Both employers and individual employees are required to comply with the requirements of the act. Enforcement and administration of the act are handled by the secretary of labor and the Occupational Safety and Health Review Commission, with some assistance from HEW.

OSHA sets out some general rules that pertain to all warehouses, manufacturing plants, and businesses. For example, passageways and aisles must be kept in good repair and maintained without obstructions. All employees involved in drilling, riveting, grinding, pouring metal, or other activities in which there is a potential hazard to the eyes must wear protective face shields or goggles. OSHA also sets specific health and safety standards for specific types of industry or business.

Both employers and employees may challenge the validity of OSHA rules in the U. S. Court of Appeals.

A worker or representative of a worker may request an inspection by OSHA. OSHA safety inspectors may enter a company without delay at any reasonable time. If an inspection reveals a violation, the employer is given a wirtten citation describing the problem. The employer than has 15 days to correct the violation or to appeal. Additional time may be granted by the U. S. Department of Labor if the time limit does not appear reasonable and the extension is requested in good faith. If a violation is not corrected, a monetary penalty of up to $10,000 may be imposed. A willful violation by an employer that results in the death of an employee is punishable by fine of $10,000. Second convictions carry heavier penalties plus a jail term.

Since one of the purposes of the law is to determine how and why injuries occur, OSHA regulations require businesses to keep certain records. These records can then be compiled on a national basis and studied to determine what corrective measures need to be taken. OSHA Form 100 is a log that must be maintained at each individual work establishment concerning any serious injury, illness, or death. OSHA Form 101 is a required supplemental record of additional details concerning deaths or injuries. OSHA Form 102 is a summary prepared annually that incorporates information from Form 100 and Form 102. An additional form, Form 103, is required of employers of more than 100 persons.

The rules and requirements set by OSHA are too extensive to list in a book such as this. Safety or security directors of companys must study those that pertain to their particular industry or business.

REVIEW AND DISCUSSION

1. Would you say that preventive approaches to safety and security are compatible? What similarities are found in the preventive aspects of safety and security programs? Explain.
2. What is the basic purpose of the Occupational Safety and Health Act (OSHA)?
3. What classes of workers are covered under OSHA?
4. Under what conditions may an inspection of a business be made by an OSHA inspector? May an employee working in a business or industrial establishment request an inspection? If so, how?
5. Are workers held responsible for following OSHA rules, or do these requirements apply only to the employer?
6. Outline what can be expected if a factory or business is found to be violating safety or health standards.

7. Are citations set up to allow a reasonable time to correct a violation noted by an OSHA inspector? How does the citation procedure work? Is there an opportunity for an appeal?
8. What penalties may be imposed if a serious violation is not corrected?
9. Outline the basic record-keeping requirements under OSHA. Of what benefit are these records to the nation as a whole?

18

Security Management and Administration

This chapter examines some of the problems related to security management and administration. The basic purpose of a security program is to provide the capacity to protect employees and visitors and to control loss. Systems and procedures that set up these capabilities come about through careful planning, management, and administration.

INTRODUCTION

Increasingly, businesses and institutions have come to appreciate the value of security. As a result, increasing attention has been directed toward specific security responsibilities and functions, including those of management.

To deal systematically and effectively with security problems, everyone concerned must have a clear understanding of where security fits and how security works in the organizational framework. It is important for managers and supervisors to realize what security can accomplish, to be aware of the techniques that can be used, and to grasp how other departments and employees of the organization are involved.

THE SECURITY DIRECTOR'S JOB

In today's highly technical and professional society, a security director or manager may find that his job spans many areas of expertise. The following are the most important functions that are handled by a security director or manager:

1. *Administrative.* Carrying out programs and policies that have been worked out with top management; setting up financial controls and budgets; educating workers as to how security procedures can benefit all employees and departments; setting up and administering training programs for company investigators, guards, or other security employees.
2. *Managerial.* Hiring or discharging security employees; scheduling work assignments; supervising and managing day-to-day functions and responsibilities of these security employees—in short, handling the management functions of the department.
3. *Investigative.* Looking into day-to-day problem situations for management; investigating applicants for sensitive company positions; looking into other personnel matters as desired by top management; handling investigations that are outgrowths of employee theft, embezzlement, or breach of company regulations; cooperating with police and fire department officials, company insurance representatives, OSHA, and other governmental inspectors.
4. *Loss Prevention.* Assuming responsibility for setting up preventive programs to protect company or institutional employees and monetary assets. This is usually called the loss prevention aspect of security.
5. *Technical.* Having knowledge of technical security problems; having at least a basic knowledge of security equipment and devices and how they function, including information as to cost-effectiveness of different security approaches, techniques that can be used, where security devices and equipment can be obtained, how installations may be made.

The director or manager of security must also establish rapport with top management and with other departments in the company in order to integrate security requirements with organizational performance and objectives. This will help the manager to plan for future contingenices and control security employees.

Examinations of company organization frequently reveal that security directors or managers are not given sufficient standing in these organizations to enable them to advise top management on a face-to-face basis; neither can they develop the support they need to work effectively.

This does not mean that a security director should have authority beyond this official's responsibilities. But it is sometimes observed that a security director does not have a direct channel to the company's owner or to management, and may be hampered in trying to inform management of facts that indicate wrongdoing or embezzlement by highly placed employees.

In a situation of this kind, the security director for a large chain of retail stores in Florida discovered facts that indicated an apparent theft by a high-ranking supervisor. Unable to report directly to top management, the security director brought the evidence to the attention of a vice-president directly over the security director in the chain of command. This vice-president, however, diluted the facts considerably in reporting the matter, since the suspected thief worked for another vice-president who was on friendly terms with the vice-president in the chain of command above the security director.

Experience shows that in order to achieve objectives, the head of security should work out an understanding within the company as to the security department's level of authority. This authority should be spelled out, not only for the benefit of the security department but so that other departments in the company understand why certain measures are necessary. Unless this is done, officials may refuse to give their whole-hearted support to security techniques and programs.

Appendix 2 contains a typical management statement granting authority to its security department.

Enlisting the Cooperation of Other Departments

Security must serve the objectives of the whole company or organization. The programs used must provide for continued protection of the organization without significant interference with essential activities. In short, security must be tailored to operate in the atmosphere in which the firm's activities are carried on rather than to conform to the ideas of an individualistic security director.

Experience shows that if the security director or manager is to be successful, a positive attitude toward problems of other departments must be demonstrated. Working with other people may be one of the most challenging problems. The security executive must not only evaluate security problems and institute corrective programs but also create a climate favorable to these programs throughout the entire company. If the security officer has not had adequate experience in establishing programs and educating employees, it is recommended that appropriate security seminars be attended.

An effective security program will usually cut across departmental lines and enter into every phase of company activity. Properly presented, a company-wide security program may remove disruptive or damaging problems that are common to all departments of the business. Unless such a program is presented diplomatically, however, some management officials may resent the program and feel that they have lost some of their authority to the security director. Situations of this kind can usually be avoided if the security director or manager has previously used a regular program of indoctrination to clearly define the role of security in benefiting all parts of the organization.

Organizing the Security Department

In most organizations, the security director or manager has the responsibility to organize and supervise day-to-day activities of the security department. This usually begins with the hiring of investigators, guards, or other security employees. Employees sought should demonstrate honesty, loyalty, and ability. The whole security organization can suffer if one employee of the department creates resentment by overstepping bounds and assuming responsibilities beyond the scope of security.

The hiring of an outside guard service may eliminate some of the administrative duties that fall upon the head of security. Even though an outside guard supplier is utilized, there must be verification of performance on a continuing basis and also supervision to make certain that responsibilities are carried out.

The security director or manager must be responsible for organizing the entire security effort. This includes setting standards for employee performance and holding employees accountable when standards are not met.

Scheduling and budgeting are also vital functions in the handling of administrative responsibilities.

In the overall security plan, the director should include a program for determining what improvements need to be made. Periodic reviews of employee performance and equipment performance will reveal whether the company needs to increase or decrease its staff, replace or repair equipment, or obtain additional technical devices. Clearly, security officials must have extensive knowledge in the use of security techniques and equipment. Finally, leadership ability is the key to effective organization, supervision, and control of the department.

The Loss Prevention Approach

For many years, a widely held belief by management was that security problems could be solved only by increasing sales or production. In recent years, many progressive managers have expressed the opinion that it is better to eliminate security problems through a preventive approach rather than to overcome problems after they occur.

In the historical pattern, managers turned to security for help only after it was obvious that a major security violation had occurred. Security was then instructed to conduct an immediate investigation.

Getting to the bottom of each incident involving a breach of security is important. But often these incidents could have been prevented by advance planning, saving the company both time and money. A preventive approach to problems, therefore, is preferable to an investigative approach. Management and the security department must work jointly in establishing programs designed to keep security violations to a minimum, isolating and correcting security weaknesses before they lead to actual incidents. The results of a preventive approach to security should not necessarily be judged by the number of shoplifters, thieves, or company violators detected but by the overall company accounting figures that represent the shrinkage or losses from stock.

If a loss prevention program is to be successful, top management must have confidence in the director or manager of security. This individual will set up appropriate controls over personnel, money, and the movement of merchandise or valuables. At the same time, these controls must be practical and interfere with other company operations as little as possible.

This does not mean that investigations will not be necessary from time to time. Every serious breach of security should be investigated to determine the responsible individuals, who can then be dealt with on an individual basis.

But from a wider viewpoint, the discovery of a loss or theft may be merely a symptom. The specific incident may indicate that systems and controls in this area either have not been set up or are not functioning properly

Justifying the Security Program

In many companies, it is necessary for the director of security to justify the programs used. Frequently, the security director could utilize the time spent in justification for more worthwhile activities. Many of the operations or activities in a business are directly involved in bringing in money. Security is not in this category; however, an effective loss prevention approach usually saves far more than such a security program costs.

Nevertheless, because of the security department's failure to bring in revenue for the company, the security director is often required to justify the department's activities. One way he can do this is to point out the value of the merchandise and money that would be left unprotected were it not for the activities of that department.

In addition, the security director can frequently show that inventory shrinkage or loss figures have been dramatically reduced through using properly applied preventive techniques.

Some security departments have justified their programs by maintaining detailed statistics on the dollar value of stolen merchandise recovered and the number of thieves or violators apprehended or detected. While these figures may be helpful in obtaining an adequate budget for the security department, they may indicate strong investigative activity rather than an effective loss prevention approach.

The director of security may find it necessary to conduct cost-effectiveness studies; in some instances, the results may be used to persuade management to authorize funds for additional projects to round out the security program.

Managing Undercover Operations Within the Company

One of the most important tools available to security management is the undercover operator. A program that makes use of undercover operators will not automatically eliminate security problems, but may be of help to management in many ways.

To function effectively, managers need to know what rank and file workers think and how they respond to various situations. Frequently, these employees have answers to specific management problems but feel that they should not communicate with supervisors or managers. If the organizational structure isolates management, business decisions will often be based on inadequate information.

The director or manager of security can frequently use an undercover operator to turn up thefts, or other violations of company interest. Managed effectively, however, the undercover operator can make much more far-ranging contributions. If properly briefed, for example, the undercover operator can discover which company programs are working and which procedures are being ignored. An alert employee of this type can also identify areas where controls should be instituted to prevent specific problems. As a caution here, the identity of the undercover operator should be disclosed to only one or two top management officials. Otherwise, disclosure is more likely.

This does not mean that the undercover employee should not be used to identify internal thieves or specific instances of misconduct. By broadening the horizons of the undercover operator, however, the security manager may be able to obtain other meaningful information as to company systems and programs that need management assistance or remedial action.

PRIVATE SECURITY PROBLEMS

Although the future of private security is promising, the industry is not without problems. Perhaps the most serious deficiencies in the private sector of security are attributable to inadequate background screening of persons hired as uniformed guards and inadequate training of these guards.

In many areas of the United States, guard contracts are awarded on a competitive basis. To make a profit as low bidder, some guard suppliers have conducted only a minimum of preemployment screening of new guards, have hired at the lowest possible wage rate, and have placed guards on the job with a minimum of training.

By far the majority of individuals seeking employment as guards are honest and reliable. But in localities where the labor market is restricted, guard suppliers receive the majority of their applications from (1) the young and inexperienced, (2) the elderly who have physical limitations and who are seeking to supplement a pension, and (3) people only marginally employable. Many of these individuals have been rejected for regular police work because of physical, educational, or background limitations.

The security industry itself early recognized the problems arising from poor hiring practices and from lack of training. To their credit, many conscientious security firms have hired carefully and have upgraded training. Often they have done this at the expense of losing guard contracts to guard suppliers who were not so concerned.

The Rand Report,[1] a study begun in 1970 and completed in 1972, focused national attention on problems and needs in the selection, quality, and training of guard personnel.

---

[1] The Rand Report was in five volumes: *Private Police in the United States: Findings and Recommendations; The Private Police Industry: Its Nature and Extent; Current Regulatory Agency Experience and Views; The Law and Private Police; Special-Purpose Public Police,* Santa Monica, California, 1972. This study was conducted by the Rand Corporation, 1700 Main Street, Santa Monica, supported by a grant from the National Institute of Law Enforcement and Criminal Justice, LEAA, U. S. Department of Justice, under the direction of James S. Kakalik and Sorrel Wildhorn. The report and survey were authorized "to describe the nature and extent of the private police industry . . . and to develop policy and statutory guidelines for improving its future operations and regulation."

In 1976, the National Advisory Committee on Criminal Justice Standards and Goals focused additional attention on the problems of guard selection and training. In the committee's 1976 publication, *Private Security: Report of the Task Force on Private Security*,[2] it was pointed out that untrained security guards are not only a waste of money but also sometimes a danger to themselves and to others.

As a result of legislation urged by the National Advisory committee on Criminal Justice Standards and Goals as well as other national groups and local associations, laws governing private security have been passed in a number of jurisdictions. These laws have varied somewhat from state to state, but they all serve as a basis for compulsory upgrading of personnel and training requirements in the guard industry.

Differences Between Sworn Police Officers and Private Security

There are times when friction does exist between public law enforcement representatives and private security.[3]

One problem that has consistently caused friction is the transmittal of false alarms by private alarm companies. Police officers frequently roll to the scene of a reported robbery or burglary, risking their personal safety. A high percentage of alarms of this kind have turned out to be false in recent years.

Valuable time is often lost through responding to a false alarm, and police personnel and equipment are in jeopardy. Police officers are justifiably disturbed by repeated incidents of this kind. As a result, some law enforcement agencies give low priority to responding to alarms. While this is understandable, a slow response may mean that a dangerous criminal is allowed to escape or that a citizen in need is seriously injured before help arrives.

Since 1975, a number of cities have passed ordinances levying a fine on the alarm company for all false alarms after the first such signal from a specific installation. Although statistics are not available, it is generally believed that local laws of this kind have caused alarm companies to improve their performance.

Security associations and groups have also pointed out to individual alarm companies that false alarms are self-defeating in the retention of existing accounts and in obtaining additional business.

There are, of course, other differences between police personnel and private security representatives. Most of these differences can be resolved, however, by dedicated people.

[2]U.S. Department of Justice, *Private Security: Report of the Task Force on Private Security*, Law Enforcement Assistance Administration (Washington, D.C.: U. S. Government Printing Office, 1977).

[3]Ibid., p. 20.

Career Opportunities in Security

In recent years, state and local police agencies have consistently made strides in learning criminal techniques, in personnel training in investigation and law, and in other aspects of education. Increasing amounts of money have been devoted to police activities. But for approximately 40 years the national rate of crime has continued upward. Increasingly, business owners and managers have turned to private security in areas where there is simply no public, or governmental, police coverage. It can be anticipated that private security will continue to expand, especially in areas where police departments are understaffed or have no jurisdiction.

Private security is a growing and profitable business field. There is virtually no aspect of society that is not affected by security in one way or another. Employing more than one million people, it is a multibillion-dollar-a-year business. For at least a decade, the private security industry has grown at a rate of 10 to 12 percent per year.

There is a real shortage of qualified individuals in security management. The dedicated professional who has training and experience should expect to be rewarded, from the standpoint of both money and job satisfaction. Opportunities should continue to proliferate, for those who are prepared and educated.

Careers in this field are limited only by the determination and energy of the individual.

The material in this book presented an introductory view of career security, touching only on the specializations that are developing throughout the private security industry.

SUMMARY

An understanding as to how security fits into the organizational framework is necessary in order to deal effectively with security problems. The interplay between security and other activities should also be understood. A successful security director may be involved in at least five kinds of activity:

1. *Administrative.* Handling programs and policies that have been worked out with top management; setting up financial controls; planning and administering training and work programs for company investigators, guards, and other security employees.
2. *Managerial.* Scheduling work assignments for security employees; hiring and discharging as needed, and supervising day-to-day activities of the security department; budgeting.

3. *Investigative.* Handling the confidential investigations needed by management; solving thefts and other crimes against the company or company officials.
4. *Loss Prevention.* Setting up programs that protect company goods and money, so that losses are not likely to occur; implementing preventive aspects or programs that have been established.
5. *Technical.* Serving as the firm's or organization's technical expert on security techniques, devices, and equipment.

The security director may be seriously hampered in handling the functions listed, unless the director is able to enlist the cooperation and support of top management and other departments in the organization. An effective security program is generally one that has been "sold" to other officials.

The preventive approach to security has gained management backing in recent years. For a long time many businessmen believed that security weaknesses in an organization could be overcome by increasing sales or production to take care of losses. Lowered profit margins and increased costs have led many officials to realize that a preventive approach usually costs far less than a corrective approach, which involves overcoming problems after they occur.

In a loss prevention approach, management and security work jointly to set up controls that will keep security losses to a minimum, thereafter auditing the controls to make sure they are still effective. The controls set up are those that regulate the movement and storage of company merchandise and money.

While there are other activities that may appear to be more significant, the director of security may be obliged to maintain statistics on merchandise recoveries and thieves apprehended by the security department. Generally used to support budget requests, this information is needed by the security department to justify its activities in some organizations.

Handling undercover operators may also be a responsibility of the security director. If possible, the identity of undercover agents should be known to only one or two members of top management. Placing and maintaining an individual in a position of this kind is always a delicate and discreet job. If given good instructions and interviewed regularly, an intelligent undercover operator may be able to pinpoint thieves, drug users, and other objectionable employees in the work force. An equally important contribution the undercover agent may make is to point out where management controls are lacking and where rules and controls are being violated. This kind of information, if properly accepted by management, provides the tools for correcting deficiencies in the management process.

In recent years, businesses and industries have come to rely on private security. The need for qualified, trained security managers and officials should expand considerably for several decades. The opportunities in this field are limited only by the determination, energy, and education of the individuals involved. At the same time, a lack of training and professionalism in the uniformed guard sector must be overcome for the good of the industry. This should come about in the near future because of required training programs. Overall, the future of the private sector of security appears bright.

REVIEW AND DISCUSSION

1. Explain the basic administrative responsibilities of a director or manager of security. Also describe the managerial and investigative responsibilities.
2. Why is it important for the director or manager of security to report directly to top management? Explain.
3. Why does the security department need the cooperation and help of the other departments and department heads in the business?
4. From an internal standpoint, what are some of the problems in organizing and supervising the security department?
5. What is meant by the so-called loss prevention approach to security? Explain the responsibilities of the security director or manager in this area. What planning may be involved here?
6. Is it sometimes necessary for the security director to justify the activities and programs of the security department? Describe the pros and cons of this kind of activity.
7. How can an undercover operator be used to the benefit of company security? Go into detail.
8. Describe some of the career opportunities in the security field in the immediate future.

Appendix 1

Specifications for Chain-Link Fencing

It may be advisable to contact city authorities to determine whether city governmental codes have placed restrictions on fence building. In some locations, the municipal code may prohibit the use of barbed wire that extends either inward or outward from the fence line. Some local ordinances specify that barbed wire topping may not be used at all. Then, too, if the fence is located exactly on the property line, it may be inadvisable to install barbed wire topping that extends into or over adjoining property areas.

A minimum overall height of at least 8 feet is considered a basic requirement for a security fence. In the typical installation, this would require 7 feet of mesh, topped with approximately 1 foot of tautly strung barbed wire. Utilizing at least 4-inch steel pipe posts at the corners, the fence bracing should be located inside the installation.

It is desirable to utilize mesh that is no larger than a 2-inch square. Unless this advice is followed, the fence may be easy to climb.

As to wire size, No. 9 gauge or heavier wire (American gauge) is recommended. Unless fence posts are set in concrete and are heavy enough to stand bumping or contact from vehicles, the usefulness of the fence may be short-lived.

Angled bars, or arms, extending both inward and outward, should be used to carry the barbed wire on top. The bars should form a "V" at an angle of about 45° from the perpendicular, with at least three strands of wire on each angled bar.

The entire fence installation should be inspected regularly, both for evidence of breaks and to make certain the barbed wire remains taut.

It is frequently observed that the bolts and nuts holding hardware attachments on a chain-link fence installation have not been welded. This oversight allows a would-be intruder to disconnect the bolts and nuts, permitting the removal of a section of mesh. The nuts can usually be removed from the bolts in a matter of a few minutes, using a pair of pliers or a small crescent wrench. Welding the nuts at the time of installation will solve the problem, or a simple ball-peen hammer may be used to flatten the threads on the bolts, accomplishing the same result.

Another commonly observed weakness in security fencing is failure to anchor the chain-link fabric properly to the ground or paving. In some installations a light cable or heavy bottom wire may be woven through the mesh, strung very taut. A mere selvedge wire at the bottom is seldom strong enough to prevent an intruder from crawling under. Preferably, the fabric should be set in cement curbing, or should be bolted to paving.

Prior to installation of the fence, a survey should be made to determine whether there are drainage ditches or culverts that will permit access under the fence. Experience shows that an opening of more than 96 square inches should be covered with a heavy metal grill or steel bars set in masonry.

FENCE GATES

The installation of a security fence is seldom effective unless the gates in the fence are properly installed and controlled. As a general requirement, the gate should be as high as the adjoining fence, and the barbed wire topping on the gate should conform to topping on the fence. If barbed wire on the gate would interfere with opening and closing, it is preferable for the topping to extend straight upward, rather than to be left off completely.

At an industrial or business location, it is difficult to install an adequate gate across a railroad track, since the gate on each side of the track must be contoured to the railroad roadbed.

It is almost always preferable to secure a single gate by using a so-called post locking arrangement. This hardware device makes use of a locking bar and latch, welded to the gate frame.

A double gate opening in a fence is frequently secured with a padlocked chain. This may be satisfactory if the chain is consistently kept tight and there is little play in the gates. If a center post installation can be used, however, better security can be expected.

If a gate can be locked from inside the premises, it is suggested that a metal shield be welded to the frame of the gate. This makes it difficult to reach the lock from outside, or to manipulate the lock.

Some business and industrial facilities make use of electrically controlled gates, especially gates used for vehicular traffic. Operated by a guard or attendant by means of a push button, these gates are sometimes consistently left open for long periods. During rush hours or during some busy shifts, it may be impractical to keep these gates closed. Better security can be expected, however, by closing the gate after the departure of each vehicle.

Securing a padlock to a vehicle gate by a short chain welded to the shackle of the padlock is a recommended method to prevent removal or substitution of the lock. Security of any gate will, of course, be compromised unless the keys to the lock are rigidly controlled.

Also, night inspections of a business or industrial fence will usually reveal that the fence is not very effective unless there is proper lighting along the fence line, inside the enclosed yard, and in gate areas.

SIGNS ON THE FENCE

Signs are advisable at regular intervals along the perimeter, stating that the property is not open to the public and that intruders will be prosecuted. This frequently has a deterrent effect, especially if the applicable city or county code section prohibiting trespassing is noted on the sign.

FENCE ALARMS

Fence alarms are to be recommended at many business and industrial installations. A variety of fence alarms are available, but a number are subject to false alarms. Blowing in a wind storm, an old newspaper may set off a fence alarm, or a stray dog leaning against the fence may accomplish the same result in some systems. Laser beam alarms, projecting along the inside of the fence, are usually effective and free from false signals but are generally quite expensive. Regardless of the type of fence alarm, it may be of little use if guards or other employees are not available to respond to alarm signals.

Appendix 2

Miscellaneous Forms

SALVAGE OR SCRAP MATERIAL SALES FORM

```
                                              #_____

                            Location_____
                                Date_____
                Store, warehouse, or branch _____


           This is to acknowledge purchase of _____
pounds or units of _____. This
material was received at _____.
Material was weighed by:_____.
and the weight was witnessed by _____.


                        _____
                              Purchaser
```

DESCRIPTION FORM TO BE FILLED OUT BY WITNESS

Description Form To Be Filled Out by

Witness

(Do not consult with other witnesses on answers)

Name of Witness_____
Date_____   Please do not guess. Leave spaces
                            blank if data were not observed.
Description of Holdup Man
Sex_____ Nationality_____ Age_____ Weight_____
Build_____ Height_____
Hair: Color_____ Curly or straight_____
      Length_____ Curly_____ Other features of hair_____

Race: Color of Skin_____ Complexion_____
      Irregularities_____ Blemishes_____
Facial Characteristics: _____
       Oval_____ Square_____ Round_____ Chin_____
Unusual Features: _____ Scars_____
       Marks or moles_____ Pock-marked_____ Other_____
Eyes & Eyebrows: Color_____ Shape_____
       Eyebrows _____ thin_____ color_____
Nose: Large_____ Small_____ wide_____ flat_____
      hooked_____ shape of nostrils_____
Mouth: Lip shape_____ large_____ thick_____ protruding_____
Teeth: Large_____ buck teeth_____ missing_____ crooked_____
       dirty_____ smooth fit_____ jaw_____
Speech: _____ Sectional accent_____
        style of speaking_____
        words actually used_____
Mustache, beard, sideburns: _____
        false_____ length_____ curly_____ color_____
        thickness_____
Cheeks: _____ pronounced_____ thin_____ sunken_____
Ears: Protruding_____ flattened_____ large_____ small_____
Neck: Large_____ thick_____ long_____ thin_____
Shoulders: stooped_____ sloping_____ broad_____ thin_____
Chest: Broad_____ flat_____ muscular_____
Hands: Covered with gloves_____ thin_____ long-fingered_____
       tattoos on hands_____ scars_____ right-handed_____
       dirty_____ fingernails_____ ring or watch_____
Clothing: hat_____ coat-style_____ jacket_____
          sweater_____ tie_____ shoes_____ socks_____
          color of coat or jacket_____
          color of pants & style_____
          color of shirt_____
          color of shoes & socks_____
Have you ever seen this person before?_____ where?_____
Weapon used_____ gun, rifle, pistol, revolver_____
Getaway car: before robbery_____ after_____
             number of persons in car_____ direction of flight_____
Disguise worn or discarded:_____ kind_____
        where discarded_____
Any other articles or evidence dropped or discarded_____
Items touched by the holdup man_____

TELEPHONE OPERATOR'S BOMB THREAT REPORT

TO BE KEPT NEAR TELEPHONE

INSTRUCTIONS: Do not hang up on the caller. Listen, be calm, be courteous. Signal to another operator that a bomb threat call is in progress (by prearranged signal).

DATE	TIME

EXACT WORDS OF PERSON PLACING CALL

ASK THE CALLER, IF POSSIBLE:

1. When is the bomb going to explode? _____
2. Where is the bomb right now? _____
3. What kind of a bomb is it? _____
4. What does it look like? _____
5. Why did you place the bomb? _____

UNDERSCORE PERTINENT INFORMATION:

Caller's Identity: Male   Female   Adult   Juvenile   Age _____ years

Voice: Loud   Soft   High Pitch   Deep   Rapsy   Pleasant   Intoxicated   Other _____

Accent: Local   Not Local   Foreign   Region _____

Speech: Fast   Slow   Distinct   Distorted   Stutter   Nasal   Slurred   Lisp

Language: Excellent   Good   Fair   Poor   Foul   Other _____

Manner: Calm   Angry   Rational   Irrational   Coherent   Incoherent   Deliberate   Emotional   Righteous   Laughing   Intoxicated

Background Noises: Office Machines   Factory Machines   Bedlam   Trains   Animals   Music   Quiet   Voices   Mixed   Airplanes   Street Traffic   Party Atmosphere

OTHER IMPRESSIONS:

IMMEDIATELY NOTIFY YOUR SUPERVISOR AND THE DIRECTOR OF SECURITY AS SOON AS THE CALLER HANGS UP.

_____ Operator

STATEMENT OF BUSINESS SECURITY

A statement of authority and responsibility recently set down by management and security for a Los Angeles firm:

It is essential that _____ Company continue its operations profitably, with the least possible loss of interruption from illegal activities or undesirable influences which may properly be avoided.

In order to maximize profits, the company has employed a director of security and special affairs. This position was instituted and developed to fill the need for a special service type department which would be available to accomplish the investigative needs of the entire company and to work with other ranking officials.

Security and loss prevention activities must necessarily be coordinated by the director of security. Preventive security as such, however, is not a separate and distinct responsibility of the security department. It is so closely aligned with good management practices as to be inseparably linked with overall managerial competency. It is therefore essential that other departments of the business work closely with the director of security.

To hold losses at a minimum, the security department will automatically enter into and investigate incidents that appear to involve possible financial loss to _____ Company and its affiliates. Any official of the company may request the services of the security department at any time the known facts warrant such a request.

The functions of the security department are not intended to supplant in any way the administrative or executive responsibilities of the various operating departments or of the individual supervisors who discharge administrative duties normally assigned to them. The security department has authority to investigate any incident or set of circumstances that indicate that _____ Company or its affiliates has incurred, or might incur, financial loss under circumstances which indicate dishonesty, thievery, or embezzlement.

The security director has authority to recommend procedures to individual departments for the control of loss. For their part, individual executives and administrators should work with the security department in installing and making effective those controls that can be expected to assure honest performance from rank and file employees.

Insofar as efficiency can be maintained and costs kept within workable limits, the recommendations by the director of security should be followed, both for actual controls and for theft deterrents.

In addition, the director of security should have authority to audit and conduct verification examinations of those procedures set up to control loss from dishonesty, embezzlement, waste, or lack of concern for company property. Upon notification, the security department should proceed as soon as expedient to conducting necessary investigations and to the auditing of security procedures. In all instances, however, it will be incumbent upon the director of security to keep top management advised of developments that reflect on the integrity of individual employees or of the control processes being used in the business.

If it becomes apparent to the director of security that company procedures have loopholes that may be used by dishonest employees, then it shall be the responsibility of the director of security to make recommendations that will eliminate the weakness in future business transactions.

When an employee is believed to have defrauded the company, the major department head concerned will be immediately advised of developments in that department. In all cases, when the matter first comes to the attention of company personnel, a security investigator of the company will, where available and in accordance with established procedures, be immediately consulted, and this individual's advice and assistance sought. Upon the arrival of the security investigator at the scene of the suspected or actual violation, the investigator will have complete charge of the investigation, subject to frequent consultation with the executive vice-president or other designated management official. The security investigator will conduct operations and proceed with investigation to ascertain whether there is a weakness in company procedures and to determine the identity of the specific individual responsible. Upon determining the existence of a weakness, the security investigator will make a recommendation for whatever changes in procedures may be warranted by the facts developed. In all inquiries and investigations the director of security will receive the active cooperation of all company personnel.

In the event a security investigator of the company is temporarily unavailable and the situation is of such urgent or emergency nature as to appear to require the immediate presence of the law, the ranking company official should have an appropriate call made to a police officer. Upon the arrival of the officer at the scene, the situation should be explained, but the actions of the officer should not be directed, nor should the officer be told what to do or how to accomplish it. Help to the officer should not be volunteered, but assistance should be given if requested to protect company personnel or property. Company personnel will not sign complaints or warrants for arrest without first obtaining the approval of the director of security.

All supervisors are responsible for fully informing their subordinates of security regulations and the resulting consequences of any violation. In addition, all supervisors should immediately report facts that indicate the possibility of a company loss to the security department. At all times the director of security has the responsibility for advising the executive vice-president of pertinent investigations or indications of procedural weaknesses within the company systems. The above is in no way intended to preclude the assignment of security investigators to other type inquiries and investigations, when so directed or requested by appropriate management officials."

Appendix 3

Model Burglary Security Code

OAKLAND
POLICE-FIRE AND INSURANCE COORDINATING COMMITTEE

MODEL BURGLARY SECURITY CODE
MINIMUM STANDARDS

I. PURPOSE

The purpose of this Code is to provide minimum standards to safeguard property and public welfare by regulating and controlling the design, construction, quality of materials, use and occupancy, location and maintenance of all buildings and structures within a city and certain equipment specifically regulated herein.

II. DEVELOPMENT OF MODEL CODE

The following City Ordinances were used as guides in developing the model code: General Ordinance No. 25, 1969, as amended, City of Indianapolis, Indiana — Section 605-3 — F211 Housing Inspection and Code Enforcement, Trenton, New Jersey — Section 23-405 of the Arlington Heights Village, Illinois, Code — Section 614.46 Chapter 3 of the Arlington County, Virginia Building Code — Section H-323.4 of the Prince George's County, Maryland Housing Code — City of Oakland, California Building Code — Burglary Prevention Ordinance, Oakland, California.

III. SCOPE

The provisions of the Code shall apply to new construction and to buildings or structures to which additions, alterations or repairs are made except as specifically provided in this Code. When additions, alterations or repairs within any 12-month period exceed 50 per cent of the replacement value of the existing building or structure, such building or structure shall be made to conform to the requirements for new buildings or structures.

IV. APPLICATIONS TO EXISTING BUILDINGS

(It is the Committee's recommendation that the Code apply only to new construction, additions, alterations or repairs. However, some cities may wish to include present structures. If so, the following paragraph may be substituted for III. above.)

All existing and future buildings in the city shall, when unattended, be so secured as to prevent unauthorized entry, in accordance with specifications for physical security of accessible openings as provided in this Code.

V. ALTERNATE MATERIALS AND METHODS OF CONSTRUCTION

The provisions of this Code are not intended to prevent the use of any material or method of construction not specifically prescribed by this Code, provided any such alternate has been approved, nor is it the intention of this Code to exclude any sound method of structural design or analysis not specifically provided for in this Code. Structural design limitations given in this Code are to be used as a guide only, and exceptions thereto may be made if substantiated by calculations or other suitable evidence prepared by a qualified person.

The enforcing authority may approve any such alternate provided he finds the proposed design is satisfactory and the material, method or work offered is, for the purpose intended, at least equivalent of that prescribed in this Code in quality, strength, effectiveness, burglary resistance, durability and safety.

VI. TESTS

Whenever there is insufficient evidence of compliance with the provisions of this Code or evidence that any material or any construction does not conform to the requirements of this Code, or in order to substantiate claims for alternate materials or methods of construction, the enforcing authority may require tests as proof of compliance to be made at the expense of the owner or his agent by an approved agency.

VII. ENFORCEMENT

The Multiple Dwelling and Private Dwelling Ordinances shall be included in the Building Code and enforced by the Building Official. The Commercial Ordinance shall be administered and enforced by the Chief of Police.

VIII. RESPONSIBILITY FOR SECURITY

The owner or his designated agent shall be responsible for compliance with the specifications set forth in this Code.

IX. VIOLATIONS AND PENALTIES

It shall be unlawful for any person, firm, or corporation to erect, construct, enlarge, alter, repair, move, improve, remove, convert or demolish, equip, use, occupy or maintain any building or structure in the city, or cause the same to be done, contrary to or in violation of any of the provisions of this Code.

Any person, firm, or corporation violating any of the provisions of this Code shall be deemed guilty of a misdemeanor and shall be punishable by a fine of not more than $500, or by imprisonment for not more than six months, or by both such fine and imprisonment.

X. APPEALS

In order to prevent or lessen unnecessary hardship or practical difficulties in exceptional cases where it is difficult or impossible to comply with the strict letter of this Code, and in order to determine the suitability of alternate materials and types of construction and to provide for reasonable interpretations of the provisions of this Code, there shall be created a Board of Examiners and Appeals (if none exist). The Board shall exercise its powers on these matters in such a way that the public welfare is secured, and substantial justice done most nearly in accord with the intent and purpose of this Code.

OAKLAND
POLICE-FIRE AND INSURANCE COORDINATING COMMITTEE
MODEL COMMERCIAL BURGLARY SECURITY ORDINANCE
MINIMUM STANDARDS

I. **ALL EXTERIOR DOORS SHALL BE SECURED AS FOLLOWS:**
  A. A single door shall be secured with either a double cylinder deadbolt or a single cylinder deadbolt without a turnpiece with a minimum throw of one inch. A hook or expanding bolt may have a throw of ¾ inch. Any deadbolt must contain hardened material to repel attempts at cutting through the bolt.
  B. On pairs of doors, the active leaf shall be secured with the type lock required for single doors in (A) above. The inactive leaf shall be equipped with flush bolts protected by hardened material with a minimum throw of ⅝ inch at head and foot. Multiple point locks, cylinder activated from the active leaf and satisfying (I, A and B) above may be used in lieu of flush bolts.
  C. Any single or pair of doors requiring locking at the bottom or top rail shall have locks with a minimum ⅝ inch throw bolt at both the top and bottom rails.
  D. Cylinders shall be so designed or protected so they cannot be gripped by pliers or other wrenching devices.
  E. Exterior sliding commercial entrances shall be secured as in (A, B & D) above with special attention given to safety regulations.
  F. Rolling overhead doors, solid overhead swinging, sliding or accordion garage-type doors shall be secured with a cylinder lock or padlock on the inside, when not otherwise controlled or locked by electric power operation. If a padlock is used, it shall be of hardened steel shackle, with minimum five pin tumber operation with non-removable key when in an unlocked position.
  G. Metal accordion grate or grill-type doors shall be equipped with metal guide track at top and bottom, and a cylinder lock and/or padlock with hardened steel shackle and minimum five pin tumber operation with non-removable key when in an unlocked position. The bottom track shall be so designed that the door cannot be lifted from the track when the door is in a locked position.
  H. Outside hinges on all exterior doors shall be provided with non-removable pins when using pin-type hinges.
  I. Doors with glass panels and doors that have glass panels adjacent to the door frame shall be secured as follows:
    1. Rated burglary resistant glass or glass-like material, or
    2. The glass shall be covered with iron bars of at least one half-inch round or 1"x¼" flat steel material, spaced not more than five inches apart, secured on the inside of the glazing, or
    3. Iron or steel grills of at least ⅛" material of 2" mesh secured on the inside of the glazing.
  J. Inswinging doors shall have rabbitted jambs.
  K. Wood doors, not of solid core construction, or with panels therein less than 1⅜" thick, shall be covered on the inside with at least 16 gauge sheet steel or its equivalent attached with screws on minimum six inch centers.
  L. Jambs for all doors shall be so constructed or protected so as

to prevent violation of the function of the strike.

M. All exterior doors, excluding front doors, shall have a minimum of 60 watt bulb over the outside of the door. Such bulb shall be protected with a vapour cover or cover of equal breaking resistant material.

II. GLASS WINDOWS:

A. Accessible rear and side windows not viewable from the street shall consist of rated burglary resistant glass or glass-like material. Fire Department approval shall be obtained on type of glazing used.

B. If the accessible side or rear window is of the openable type it shall be secured on the inside with a locking device capable of withstanding a force of 300 pounds applied in any direction.

C. Louvered windows shall not be used within eight feet of ground level, adjacent structures or fire escapes.

D. Outside hinges on all accessible side and rear glass windows shall be provided with non-removable pins. If the hinge screws are accessible, the screws shall be of the non-removable type.

III. ACCESSIBLE TRANSOMS:

All exterior transoms exceeding 8"x12" on the side and rear of any building or premises used for business purposes shall be protected by one of the following:
1. Rated burglary resistant glass or glass-like material, or
2. Outside iron bars of at least ½" round or 1" x ¼" flat steel material, spaced no more than 5" apart, or
3. Outside iron or steel grills of at least ⅛" material but not more than 2" mesh.
4. The window barrier shall be secured with rounded head flush bolts on the outside.

IV. ROOF OPENINGS:

A. All glass skylights on the roof of any building or premises used for business purposes shall be provided with:
1. Rated burglary resistant glass or glass-like material meeting Code requirements, or
2. Iron bars of at least ½" round or 1" x ¼" flat steel material under the skylight and securely fastened, or
3. A steel grill of at least ⅛" material of 2" mesh under the skylight and securely fastened.

B. All hatchway openings on the roof of any building or premises used for business purposes shall be secured as follows:
1. If the hatchway is of wooden material; it shall be covered on the inside with at least 16 gauge sheet steel or its equivalent attached with screws.
2. The hatchway shall be secured from the inside with a slide bar or slide bolts. The use of crossbar or padlock must be approved by the Fire Marshal.
3. Outside hinges on all hatchway openings shall be provided with non-removable pins when using pin-type hinges.

C. All air duct or air vent openings exceeding 8" x 12" on the roof or exterior walls of any building or premise used for business purposes shall be secured by covering the same with either of the following:
1. Iron bars of at least ½" round or 1" x ¼" flat steel material, spaced no more than 5" apart and securely fastened, or
2. A steel grill of at least ⅛" material of 2" mesh and securely fastened.
3. If the barrier is on the outside, it shall be secured with rounded head flush bolts on the outside.

V. SPECIAL SECURITY MEASURES:

A. Safes:
Commercial establishments having $1,000 or more in cash on the premises after closing hours shall lock such money in a Class "E" safe after closing hours.

B. Office Buildings (Multiple occupancy)
All entrance doors to individual office suites shall have a deadbolt lock with a minimum one inch throw bolt which can be opened from the inside.

VI. INTRUSION DETECTION DEVICES:

A. If it is determined by the enforcing authority of this ordinance that the security measures and locking devices described in this ordinance do not adequately secure the building, he may require the installation and maintenance of an intrusion detection device (Burglar Alarm System).

B. Establishments having specific type inventories shall be protected by the following type alarm service:
1. Silent Alarm—Central Station—Supervised Service
   a. Jewelry store — Mfg., wholesale, and retail
   b. Guns and ammo shops
   c. Wholesale liquor
   d. Wholesale Tobacco
   e. Wholesale drugs
   f. Fur stores
2. Silent Alarm
   a. Liquor stores
   b. Pawn shops
   c. Electronic equipment
   d. Wig stores
   e. Clothing (new)
   f. Coins and stamps
   g. Industrial tool supply houses
   h. Camera stores
   i. Precious metal storage facility
3. Local Alarm (Bell outside premise)
   a. Antique dealers
   b. Art galleries
   c. Service stations

VII. EXCEPTIONS:

No portion of this Code shall supercede any local, state, or Federal laws, regulations, or codes dealing with the life-safety factor.

Enforcement of this ordinance should be developed with the cooperation of the local fire authority to avoid possible conflict with fire laws.

Index

Access controls, 98
Alarm sensoring systems, 84
Alarm system, assessing need for, 82
Alarms, 80
American Society for Industrial Security, 6
Applicant interviews, 189
Applicant investigations, 190
Appropriation of goods by employees, 121
Armed robbery, 142–62
  announcing loss, 160
  distinguished from burglary, 142
  protecting evidence at scene, 156
  protective measures beforehand, 145, 148
  recording description of subject, 159
  recording witnesses, 159
Armored car service, 10
Arrests by security guards, 90
Art and decoration protection, 240
ASIS, see American Society for Industrial Security

Assessing need for alarm systems, 82
Audits to prevent theft or embezzlement, 120
Automatic telephone dialers, 83
Automobile supplies, 64

Backup records, 214
Bad checks, 236
Badge systems, 104
"Bait money," 157
Baker Industries, 6
Bar and hammer attack on locks, 38, 39
Basement access, 30
"Beaver 55," 7
"Bell-ringer" alarms, 83
Bleeding the cash register, 149
Blind receiving, 176
Bomb threat report form (telephonic), 287
Bomb threats, 231
Bonding, 200
Break-resistant glass, 28

295

296 Index

Bugging devices, 13
Building evacuation, 232
Building placement for security, 17
Building tours, 111
Burglary, 163–171
   basic types of, 164
   conditions behind increase in, 164
   defined, 163
   limiting time and money available, 166
   losses to business, 165
   techniques, 169
   code model (Oakland ordinance), 290
Business burglary, 163
Business machine thefts, 245
Business espionage, 220

Capacitance alarms, 86
Carbon dioxide (CO_2) fire protection, 211
Cardkey access systems, 109
Cardkey locks, 44
Career opportunities in security, 279
Cash drawers, 58
Cash limits in business, 149
CCTV, see Closed circuit television
Central alarm station, 10
Central station system, 83
Chain-link fencing, 19
Check losses in business, 236
Checks, protecting in event of robbery, 150
Classes of fires, 72
Closed circuit television, 87, 88, 103
Collection losses by robbery, 126
College campus security problems, 256
Computer destruction, 7
Computer fire protection
   water sprinkler systems, 210
   carbon dioxide (CO_2), 211
   halon 1301 (freon), 212
Computer security, 204–217
   fire as major threat, 207
   location for protection, 205
   sabotage, 7, 205

Controls over theft, 119
Corporate valuables, protection of, 239
Cylinder pulling, 36, 37

Day latch for vault, 55
Delivery truck thefts, 178
Delivery vehicle safe, 54
Description form for suspected armed robber, 286
Description of armed robber, as investigative aid, 154
Detectors in computer installation, 209
Discharge for theft, 196
Disposal of computer printout, 215
Dock security, 173
Document protection, 224
Door hinge pins, 24, 25, 26, 27
Door pinning instead of locks, 23
Door protection, 20, 24, 25, 29
Driver collection losses, 126
Drop safe, 54
Dry pipe system (type of water sprinkler), 71
Dumpster for trash, 67, 68
Duplication of keys as security hazard, 47

Early employee, as potential thief, 125
Educational programs to reduce shiplifting, 139
Electromechanical devices, 84
Emergency exit alarms, 22
Emergency shutdown and recovery for computers, 216
Employee appropriation as form of theft, 121
Employee lockers, 112
Employee motivation for honesty, 192
Employee purchases, 123
Employee security awareness sessions, 193
Employees, exit interviews with, 199
Equipment protection, 62
Espionage in business, 220

Evacuation of building for bomb threat, 232
Executive protection, 12
Exit interviews, 199

Fence alarms, 86
Fencing, as perimeter protection, 19, 282–284
File cabinet security, 56, 57
Fire, four classes or types of, 72
Fire detectors, 76
  in computer rooms, 209
Fire doors, 75
Fire escapes, 76
Fire exit doors, 21
Fire extinguishers, 73–75
Fire hoses, 75
Fire as major threat in computer protection, 207
Fire protection programs, 70
Fire protection surveys, 13
Floor pullers, 213
Forklift as security risk, 66
Freon, 212

Gasoline losses, 64
Glass cutter entries, 28
Good general sources on security, 15
Grid systems, in fire detection, 209
Guard assignments in case of fire, 77
Guard dogs, 11, 94
Guard services, general description of, 8
Guard shack, 103
Guards, see Security guards

Halon 1301 (freon), 212
Hand extinguishers, in computer room, 213
Hand trucks, 65
"Hardening the target" against burglary, 167
Hasp protection, 41
Hinge pins, 24–27
Hospital security, 250

Incoming checks, 244
Industrial espionage, 219
Inside theft, 118
Institutional security, 249–260
Insulated file cabinets, 56
Interviews with applicants, 189
Intrusion alarms, benefits from using, 81
Investigations of applicants, 190
Investigative services for management, 9
Iron bars, compared to steel mesh, 28

Jamb peeling, 38
Janitorial access, as security risk, 60

Key control, 46–50
Key control record systems, 48–50
Key duplication, as security risk, 47
Kidnapping, in connection with robbery, 148
"Kidnapping the safe," as burglary technique, 53
Kidnappings, 7
Kinds of security, 2

Ladders as security risk, 31
Landscaping, 17
Library security, 258
Lie detector, see Polygraph examinations
Limiting time and loot available to burglars, 166
Local alarm system, 83
Location of computer for protection, 205
Location of money safe, 51
Locks and keys, 35–50
Loss prevention surveys, 11

Mailroom security, 243
Maintenance access, as security risk, 60
Marking company tools, 63

Merchandise handling, security risks, 173–187
Merchandise substitution, 127
Model burglary code (Oakland ordinance), 290
Motion detection alarms, 85
Motivating employees for honesty, 192
Museum security, 258

Neighborhood location for security, 17
Night lighting, 100

Oakland, California model burglary ordinance, 290
Occupational Safety and Health Act, 261–270
   administration of act, 263
   extensive number of standards, 267
   record keeping requirements, 266
   time for correction of hazards, 265
   violations, 264
Office art and decorations, as security risk, 240
Office security, 236–248
Office supply theft and pilferage, 241
Off-site storage of computer backup, 214
Opening time holdups, 151
Opportunities in security profession, 279
OSHA, abbreviation for Occupational Safety and Health Act
Outside theft, 118
Overshipments, 177

Package passes, 252
Padlock substitution, 40
Pallet losses, 65
Park security, 258
Parking lot problems, 99
Passenger screening, 7
Patrol services, 8, 94
Patient theft in hospitals, 253
Patrol vehicles, 100
Payroll fraud, 246

Perimeter controls, 18
Personnel security, 188–203
Petty cash, 242
Photoelectric alarms, 85
Physical controls, as deterrents to entry, 16
Physical security, three lines of defense, 17
"Pinning" doors in lieu of using locks, 23
Polygraph examinations (lie detector), 12, 191
Predeparture screening, 11
Preloading trucks as security risk, 180
Press statements after armed robbery, 159
Pressure alarms, 86
Private security, 2, 277
Problem employees, 196
Problems in private security, 277
Proprietary systems, 83
Prosecution of shoplifters, 135
Protecting checks in armed robbery, 150
Protection services, 88
Proximity alarms, 86
Protective features of safes, 56
PSE, see Psychological stress evaluator
Psychological stress evaluator, 13, 191
Public security, 2
Public utility protection, 31

Railroad seals, 182
Receiving, as security risk, 175, 177
Receiving "blind," 176
Recording lock systems, 43
Removable core locks, 41, 42
Replacement tools, 63
Residential burglary, 163
Returned merchandise controls, 184
Robbery, 142–162
   getting description of bandit, 154
Roof entry, 23, 30, 31
Roof hatch protection, 30
Route collection losses, 126
Rules for company security, 194

Safe combination protection, 168
Safes, 50–55
  protective features of, 56
Safety, responsibility for programs, 69
Salvage sales form, 285
Salvage thefts, 122
Samples, 122
Sawing lock bolt, 39
School security, 254
Scrap material sales form, 285
Scrap and salvage thefts, 122
Seals as security controls, 182
Security, 1–20
  defined, 1
  history of, 1–8
Screening of passengers, 7
Security awareness sessions, 193
Security guards, 89–90
  power to arrest, 90
  problems in carrying weapons, 91
  pros and cons, 89
  responsibilities and limitations, 89
Security rules, 194
Security management, 271–281
  basic functions of, 272
  enlisting cooperation of other departments, 273
  organizing the security department, 274
Security manual material, 287
Selecting and checking shipments, 183
Sensing systems for alarms, 84
Setting cash limits in business, 149
Shipping by UPS as security risk, 178
Shoplifting, 131–139
  age of persons involved, 132
  apprehension of violators, 135
  clothing store problems, 136
  control by building design, 132
  customer service as factor in control, 134
  educational programs for high schools, 139
  how perpetrated, 133
  prosecution, 135
  protection techniques, 134
  sex of perpetrators, 132

Shopping services, 12
Skylight window protection, 30
Side product theft, 122
Silent alarms in robbery, 152
"Smash and grab" theft, 28
Sonic alarm systems, 85
Springing door technique in burglary, 36, 37
Statements to press after robbery, 159
Steel mesh, 28
Strikebreakers, use of guards as, 5
Suppliers' badges, 106
Supply losses, 241
Surveillance of company truck, 181
Sweeps for bugging devices, 13

"Tail gate" thefts, 178
Taking merchandise samples, 122
Tape or foil and contacts, 84
Tape library controls, 215
Techniques used by burglars, 169
Telephone dialers, 83
Telephone misuse, as security risk, 246
Telephone operator bomb threat report form, 287
Telephone taps, 225
Telephone threats, 231
Temporary badges, 105
Temporary employees, 197
Theft, 116–119
  causes of, 118
  consequences of in business, 117
  controls to minimize, 119
  discharge as penalty for, 196
  extent of, 116
  from delivery truck, 178
  of entire load, 179
  in money and goods, 117
  problems inside and outside compared, 118
  security risks, 116
  of tools, 62
Ticket switching, 127
Till-tapping, 128
Time lock systems, 43, 45

Tool and equipment protection, 62–64
Tours, as security risk, 111
Trade secret information, how developed, 223
Trade secret protection, 219–230
Trash disposal problems, 67–69
Truck surveillance, 181
Truck theft, 179
Typewriter theft, 245

Undercover operations, 10, 276
Uniforms, losses of, 67

Vaults, 55–56
Vehicle safes, to protect collections, 54
Ventilation in buildings, as security risk, 29
Vibration detectors, 86
Visitor badges, 105
Vendors' badges, 106

Walk-in vault, 55
Warehouse door protection, 24
Watchclock systems, 92
Water sprinkler damage in computer installations, 210
Water sprinkler systems, 71
Weapons, risk in allowing guards to carry, 91
Welding equipment, 54, 65
Wells Fargo, 6, 10
Wet pipe system (type of water sprinkler), 71
Wheels on safe, 53
Will call counter thefts, 124
Window bars, 28
Window mesh protection, 28
Window protective features, 20, 28, 29
Wire tapping, 13
Witnesses to armed robbery, 159
Women in security, 14